Innovation in Smart Materials and Structural Health Monitoring for Composite Applications

by
F. Mustapha
A. Hamdan
Nisreen N. Ali Al-Adnani
K.D. Mohd Aris

Structural health monitoring (SHM) is an automated approach to determine any changes in the integrity of a mechanical system. The SHM system gives information in real time and online.

The knowledge on Root Mean Square Deviation (RMSD) techniques is employed and presented in this writing. Besides that, SHM system in wind turbine system is becoming very important. This book places emphasize on the application of biocomposite turbine blades for vertical axis wind turbines. The dynamics characterization of mechanical system on biocomposite turbine blades is determined with several techniques. The SHM for biocomposite turbine blades is enhanced in order for it to become a micro energy harvester.

Innovation in Smart Materials and Structural Health Monitoring for Composite Applications

written by

F. Mustapha

A. Hamdan

Nisreen N. Ali Al-Adnani

K.D. Mohd Aris

Published by **Materials Research Forum LLC**
Millersville, PA 17551, USA

Published as part of the book series
Materials Research Foundations
Volume 13 (2017)
ISSN 2471-8890 (Print)
ISSN 2471-8904 (Online)

Print ISBN 978-1-945291-28-9
ePDF ISBN 978-1-945291-29-6

This book contains information obtained from authentic and highly regarded sources. Reasonable efforts have been made to publish reliable data and information, but the author and publisher cannot assume responsibility for the validity of all materials or the consequences of their use. The authors and publishers have attempted to trace the copyright holders of all material reproduced in this publication and apologize to copyright holders if permission to publish in this form has not been obtained. If any copyright material has not been acknowledged please write and let us know so we may rectify in any future reprint.

Distributed worldwide by

Materials Research Forum LLC
105 Springdale Lane
Millersville, PA 17551
USA
http://www.mrforum.com

Manufactured in the United State of America
10 9 8 7 6 5 4 3 2 1

Table of Contents

Preface

Structural health monitoring (SHM) is an automated approach to determine any changes in the integrity of a mechanical system. The SHM system gives information in real time and online. Hence it provides advantages in damage detection, damage localization, damage assessment, and life prediction as compared to Non-destructive test (NDT) which is conducted offline. The research on SHM especially on aerospace structures has already undergone intensive exploration. Several issues exist for modern aerospace structures such as detectability of damage in composite structure, proper selection of sensors and anisotropic characteristic of composite aircraft structures. There are also other fields which apply SHM such as civil, architecture, mechanical system and marine. Composite structures are applied in civil and architecture industry as well during recent decades.

The knowledge on Root Mean Square Deviation (RMSD) techniques is employed and presented in this writing. Besides that, SHM system in wind turbine system is becoming very important. This book places emphasize on the application of biocomposite turbine blades for vertical axis wind turbines. The dynamics characterization of mechanical system on biocomposite turbine blades is determined with several techniques. The SHM for biocomposite turbine blades is enhanced in order for it to become a micro energy harvester.

Chapter 1

Smart material and structural health monitoring in composite applications - an innovative approaches in non destructive testing

F. Mustapha

Department of Aerospace Engineering, Universiti Putra Malaysia, 43400 Serdang, Selangor, Malaysia

Keywords

Composite Materials, Non Destructive Testing (NDT), Structural Health Monitoring (SHM)

Abstract

A series of advances in improving structural integrity and lifespan has been a major focuse area for critical assembly structures such as aircraft and other related aerospace systems. It is well known that the adoption of new materials and technologies into aerospace structures is very conservative and heavily dependent upon past design technology. Most modern and existing aircrafts are currently employing two design methodologies for the enhancement of the life management process, which are the *safe-life, and damage-tolerant* philosophy. Based on these procedures, the methodology is only beneficial if the critical damage location is known beforehand and moreover the drawback from this principle is that, it is a manpower related inspection which will have a significant impact on the overall direct operational costs (DOC). Even though this methodology is widely used by many aircraft designers in order to ensure safety and reliability, the frequent cycles and elaborate inspection procedures can sometimes be considered unnecessary, inefficient and can potentially lead to economic turmoil and increase the aircraft downtime. Thus, the method of inspection is critical to distinguish the structure life and true condition of the tested structure effectively. Nevertheless, selecting an inspection method is still subjective and in most cases will depend on several other factors and professional experience. The urgent need for fast and efficient selection of Non Destructive Evaluation\ Testing (NDE/T) is vital for reducing the need for manpower and aircraft down-time costs. In addition, consistent monitoring of the structural states of the aircraft structure is important for ensuring reliability and durability of the air vehicles. The process of continuous or consistent monitoring using the above NDE/T methods also known as a Structural Health Monitoring (SHM) system.

Contents

1. Introduction

The rapid growth of airplane commuters has prompted aircraft manufacturers to produce low-cost aircraft complying with safety regulations (airworthiness regulations), with improved lifespan durability (high endurance, lower structural weight, high strength, higher damage tolerance and higher corrosion resistance), environmental friendliness (reduced emission of CO_2 to the atmosphere), low cost maintainability and ultimately to produce large capacity aircrafts. In order to fulfil these requirements, one of the focused areas in the design of aerospace structures is the development of lightweight and high strength materials.

Figure 1: Composite structure for Boeing 777 [1].

Apart from the aluminium family materials used in the major part of civil aerospace applications composites have becoming significantly important and nearly 15 % of civilian aircraft and 50 % of helicopters and fighter aircraft lower their structural weight by utilizing these materials [1]. This is due to their lightweight, high mass specific stiffness and strength, and the potential for additional functionality (smart monitoring applications). According to an EADS report [1], by defining the anisotropic behaviour, the integration between sensors and actuators is also possible. Figure 1 illustrates the current status of composite used in the Boeing 777 and Figure 2 shows the development phase of composite materials adapted by Airbus and Boeing.

Share of Composite
Components

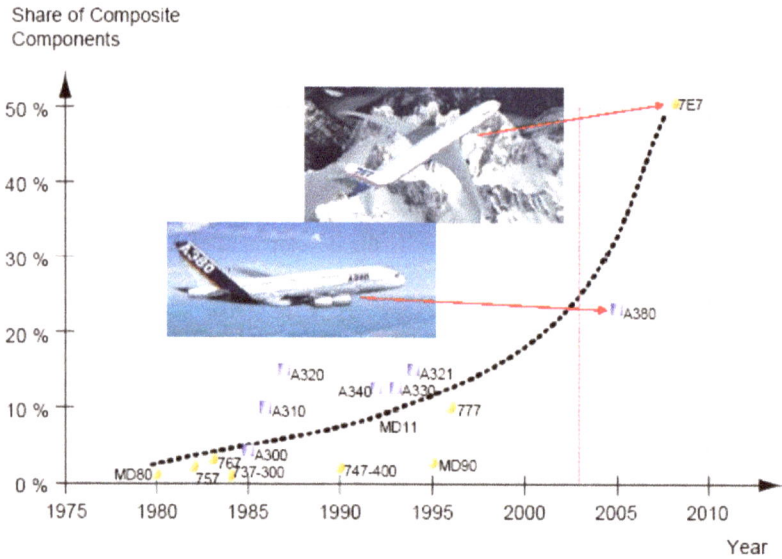

Figure 2: Showing the composite structure in the aircraft manufactures [1].

According to Diamanti *et.al.*[11], a lot of effort has been put in to finding the most reliable NDE technique for detecting defects in composite materials. Furthermore, the cost of inspection for composite structures is very high, at least one order of magnitude greater than metallic parts as highlighted by Bar-Cohen [12].The treatment in terms of design and servicing of composite structural components is drastically different from metals; variables such as labour intensive manufacturing, expensive raw materials, damage tolerant aspects, and the need for new techniques in inspection and repair philosophies need to be taken into consideration. In summary, with recent developments in the introduction of new materials (composites introduce inhomogeneous and anisotropic behaviour that is still difficult to understand) there will be a great challenge in the fields of inspection and maintenance for this type of structural assembly.

Moreover, with recent development of parts made from various materials with complex geometries, together with new manufacturing technologies (such as super plastic forming and diffusion bonding of titanium and aluminium metal sheets [13]) there will certainly be a greater challenge in the inspection and maintenance of the next generation of modern aircrafts. This occurrence will ultimately have a direct impact especially on the Direct

Operating Costs (DOC) for civil aircraft, where it is currently reported that nearly 15 % of the distribution share is coming from maintenance activities (see Figure 3) [1]. Undoubtedly, if no immediate effort is made in deriving the most reliable, cost effective, efficient and systematic Structural Health Monitoring (SHM) systems, the distribution of the DOC for these activities will consume a higher percentage, and the consequences will lead to additional costs to the airline operators and manufacturers, and unavoidably the aircraft commuters as well.

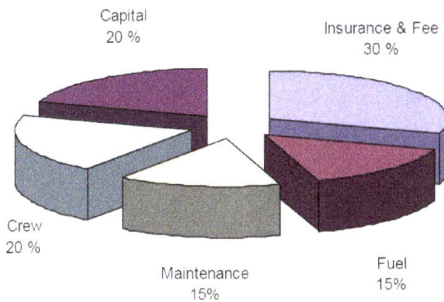

Figure 3: Direct Operating Costs (DOC) for civil aircraft [1]

2. Smart materials technology in damage detection

Inevitably, the most pressing need for this ageing aircraft problem is to embed a sensory system primarily used for monitoring the progress of the damage levels. A fully automated preferably onboard system proposed with the incorporation of smart materials technology is most welcome. According to reference [17], futuristic aircraft may be linked with the integration of smart actuators and sensors. A review of this smart actuator technology can be found in [17, 18]. One of the basic definitions of how to characterise a *smart* material is 'one or more properties of the structure can be dramatically altered' [18]. This technology is becoming a vital factor in engineering applications as lifetime efficiency and improved reliability can be gained from it [19]. Examples of smart materials that are widely used in engineering applications are piezoelectric materials, shape memory alloys (SMAs), electro rheological materials, electrostrictive materials and others.

The true design philosophy of smart structures in damage detection is the incorporationof sensors, actuators, controllers and signal processors [15]. The primary objective of this so-called smart-structure is to be sensitive to the structural degradation and the reliability of evaluating damage detection performance. The incorporation of intelligent data processing techniques such as pattern recognition, data fusion, artificial neural networks, novelty detectors and much more has assisted in optimising the results of damage identification problems [15]. It is most unlikely that a complete automated monitoring system with embedded smart structures will see its functionality for many years but this adaptive smart structure should be implemented to be an integral part in the design wheel of an aircraft design subject matters when one wants to plan for a new type of 'smart aircraft'. With regard to the ageing aircraft, the reliable automated and adaptive smart structure is considered to be a big opportunity as this technology can be seen as an added value in terms of diagnosing the structural condition with greater confidence. But, the question arising here is how to retrofit or embed this intelligent structure onto the ageing aircraft and this problem should be the focused area explored and resolved immediately.

3.   Structural health monitoring system

Structural Health Monitoring (SHM) can effectively be the problem solver for the continuous structural monitoring devices. According to Kessler [6], structural health monitoring essentially involves the embedding of an NDE system into a structure to allow continuous remote monitoring for damage detection. Hall [20] defined SHM as the "acquisition, validation and analysis of technical data to facilitate life-cycle management decisions". SHM also gives a system the ability to detect and interpret unfavourable "changes" in a structure in order to improve reliability and reduce life-cycle costs. In addition, according to [21], the aerospace industry has one of the highest payoffs for SHM since damage can cause catastrophic failures and most air-borne vehicles require regular inspections, which are very costly. It was estimated that nearly 27 % of an average aircraft's cycle expenses are due to inspection and repair [20].

The implementation of SHM systems in many industries, ranging from industrial machinery to aerospace have proven their usefulness toward signalling different structural conditions on any assembly parts or models under probe. Some of these systems are executed in-situ, one of the examples is of a thin-walled structure using embedded-ultrasonic structural radar as demonstrated in reference [22], advantages such as very good detection accuracy can be observed by adopting this method of structural monitoring. Moreover, by adapting in-situ structural health monitoring devices on any structural system, the overall system and its subsystem can remain in operation without regularly undergoing scheduled maintenance. The integrated SHM devices will systematically report when a repair is necessary, thus a planned servicing on the designated structural system or its subsystem can be executed. This planned and systematic maintenance effort can significantly reduce the overall operating cost. Currently, the structural health monitoring and damage detection agenda is the major subject of international and collaborative research effort. Van der Auweraer and Peeters [23] report significant research efforts have been made involving multi-institutions in solving the problems of health monitoring and damage detection.

As can be seen in Table 1 many collaborating efforts on this system concentrate more on aerospace structures.

Acronym-Prime Duration	Title	Structures	Key Topic/Result
OSMOS Bertin Et Cle '92-95'	Optical fibre sensing system for monitoring of structures.	Aerospace composites civil.	Quasi-distributed polarimetric sensing system; microwave sensing system; impact sensor
MADAVIC U.Napoli Federico II '96-'98	Magnetorestrictive actuators for damage analysis and vibration control.	Aerospace.	Magnetorestrictive actuators for vibration-based damage detection and active control.
FOSMET BICC Cables Ltd '96-'99	Fibre optic strain monitoring at elevated temperatures.	Power plants.	Optical fibre sensor measuring multiple points on a single strand; engineering and robustness aspects.
MONITOR BAE Defense Ltd '96-'99	Monitoring on-line integrated technologies for operational reliability.	Aerospace metallic and composites.	Multiple sensor technologies, optical fibres, acoustic emission, were evaluated and integrated in a flying test-bed
DAMASCOS Uni..Strathclyde '98-'01	Damage assessment in smart composite materials.	Aerospace.	Ultrasonic excitation combined with piezo and optical detectors and DSP analysis of scatter patterns.
MILLENNIUM Protedel Int.Ltd '98-'01	Monitoring of large civil engineering structures.	Concrete.	On-line strain measurement system based on optical fibres.

Table 1: Project Collaboration on Structural Health Monitoring for various structures [23].

Table 1 highlights national and international projects dealing with health monitoring on these structures. Smart partnerships between competent European and American experts in investigating promising techniques as a solution for different NDE aspects and to merge some of the research activities is presented and cited in [24, 15]. MONITOR (Monitoring On-line Integrated Technologies for Operational Reliability) brought together major European aircraft manufacturers with researchers and academicians in sourcing the tools by which damage detection or the prognosis of impending damage can be given to aircraft structures [15].

4.   Non destructive evaluation/testing (NDE/T) methods

4.1   Non Destructive testing (NDT)

Techniques of detecting and monitoring for materials and structural damages are based on Nondestructive Testing (NDE) [28]. Specifically, the aim is to identify and to locate physical features, which are unacceptable without causing any material damage to the material structure or the component under probing [29]. Generally, NDT is a branch of the materials sciences that is concerned with all aspects of the uniformity, quality and serviceability of materials and structures [30]. The focused of NDT incorporates all the technology for measurement and detection of significant properties, including discontinuities detection and material characterization in various engineering fields [30].

The importance of NDT can be described as follows:

i) Safety Assurance.

ii) Sensible Maintenance Schedule.

iii) Extended Service Life

iv) Increased Profit.

Compatibility between the NDT techniques and the applications (objectives) is essential in order to get optimum results with minimum cost. Numerous literatures and reports highlighting the scope of NDT and its applications are widely available. Table 2 [30] summarizes information about NDT methods. Ultrasonic, radiographic, thermographic, electromagnetic, and optical methods are most commonly employed in NDT methods.

Basic Categories	Objectives
Mechanical and optical	color, cracks, dimensions, film thickness, gaging, reflectivity, strain distribution and magnitude, surface finish, surface flaws, through-cracks
Penetrating radiation	cracks, density and chemistry variations, elemental distribution, foreign objects, inclusions, microporosity, misalignment, missing parts, segregation, service degradation, shrinkage, thickness, voids
Electromagnetic and electronic	alloy content, anisotropy, cavities, cold work, local strain, hardness, composition, contamination, corrosion, cracks, crack depth, crystal structure, electrical and thermal conductivities, flakes, heat treatment, hot tears, inclusions, ion concentrations, laps, lattice strain.
Sonic and ultrasonic	crack initiation and propagation, cracks, voids, damping factor, degree of cure, degree of impregnation, degree of sintering, delaminations, density, dimensions, elastic moduli, grain size, inclusions, mechanical degradation, misalignment, porosity, radiation degradation, structure of composites, surface stress, tensile, shear and compressive strength, disbonds, wear
Thermal and infrared	bonding, composition, emissivity, heat contours, plating thickness, porosity, reflectivity, stress, thermal conductivity, thickness, voids

Table 2 : Non Destructive Testing (NDT)

4.2   Structural health monitoring (SHM) systems and innovative approaches in NDT

The primary goal of SHM is to be able to replace the current NDE/T techniques with a continuous monitoring system [21] or to have an integrated built-in inspection system [2]. SHM is a broad field of engineering science that is composed of a number of technology interactions (sensor, signal processing, etc.). The main objective of this integrated system is to detect or to identify any possible damage present within a structure or in other words, is to monitor the health status of the structure. SHM activities have been the most focused area in critical engineering applications such as complex aircraft structures [4], civil infrastructures [67], and other mechanical structures [68]. Modal parameters, such as natural frequencies, damping ratio, and mode shapes are the most popular approach in structural health monitoring as collated in references [69-71].

These SHM techniques are mostly governed by vibration data in order to diagnose the condition of the structure under probe. Numerous methods based on vibration data characteristics for structural health monitoring were cited and reviewed by Doebling *et.al* [72]. More recently the methods of modal response and wave propagation techniques in detecting damage on an aluminium plate and beams were introduced by Mal *et al.*[73]. As for this work, the dominant SHM systems philosophy presented will be concerned with the latter technique. Farrar [74] delineated that vibration based structural health monitoring is fundamentally one of statistical pattern recognition. This process is composed of several pivotal disciplines namely; the operation evaluation, data acquisition and signal processing, feature extraction and statistical approaches.

Even though these four crucial portions are relevant to vibration-based methods, the principles behind these points can also be applied to wave propagation techniques as both of the techniques implement similar damage detection and identification strategies. On the other hand, before any of these four vital SHM points can be explored thoroughly, the understanding of how to perform systematic SHM system components and the required variables needed, are of a high priority to be investigated in a well understood manner.

In brief, the main aim of this section is to outline the general SHM research methodologies, which is basically motivated by the statistical pattern recognition process variables and the SHM system components discussed in the following section.

4.3   SHM system components

The flowchart in Figure 4 summarizes the SHM system components introduced by Kessler in his PhD thesis [6]. Only a brief description is presented here and one may refer to reference [6] for further clarifications. The integration of the system components is the key point to demonstrate whether the entire assembly system is adopting in-situ monitoring techniques. The architecture of the SHM system components can be defined as the preliminary design requirements for the efficiency of the overall SHM system [6].

The procedures of how to integrate the monitoring devices that correspond to the type of the anticipated damage, together with the loading effect of the devices on the monitored components [6], are the typical activities involved in this stage. The decision on which of the methods of monitoring one should use, such as the real-time or discontinuous systems must also be resolved in this section. Real-time monitoring is a non-stop activity in evaluating the current health status of the structural elements under probes. Whereas, the discontinuous monitoring refers to uncomplicated inspection operation and can be done periodically [6].

The second stage of the SHM system is to address the damage features that need to be evaluated. This includes studying the effect of the overall structural integration if the predicted damage is to be detected. Furthermore, the main elements in this section are to fully comprehend what are the most common types of damage occurring on the monitored structures and the consequences of the detected damage to its neighboring parts (e.g. helicopter hub and the connected shafts to the turbine blade, wing holders parts to the wing and fuselage structure). Once the damage characteristics have been established, the selection of the type of sensing method and sensors can be focused.

The inclusion of smart sensors and actuators combined with advanced signal processing techniques provides an interesting platform for cost-effective continuous monitoring devices [15].

In addition, the incorporation or the embodiment of this structure onto the tested structure is also the key issue to be considered at this stage. Sensors in SHM systems components can be categorized as passive or active. A simple analogy describing these two types of sensors is that, passive just listen to the structure whereas active can talk and listen to the structure under probe. The fourth stage of SHM components is a problem related to on how to transfer the acquired data from the sensor or actuator to/from the appropriate processors and the storage medium (computation) so as to diagnose structural condition. There are again two obvious transfer devices that can be thought of, namely the wired and wireless communications. And finally, powerful yet simple algorithms are needed to resolve the true condition of the structure under probe. The proposed algorithms must be able to interact effectively with the noisy acquired data from the sensors, and yet be compatible with the hardware configurations and software implementations in the monitoring systems. By considering the above general SHM system components, reliable and effective conduct of the developed SHM system for this work can be planned strategically and systematically so as to achieve efficient SHM operation.

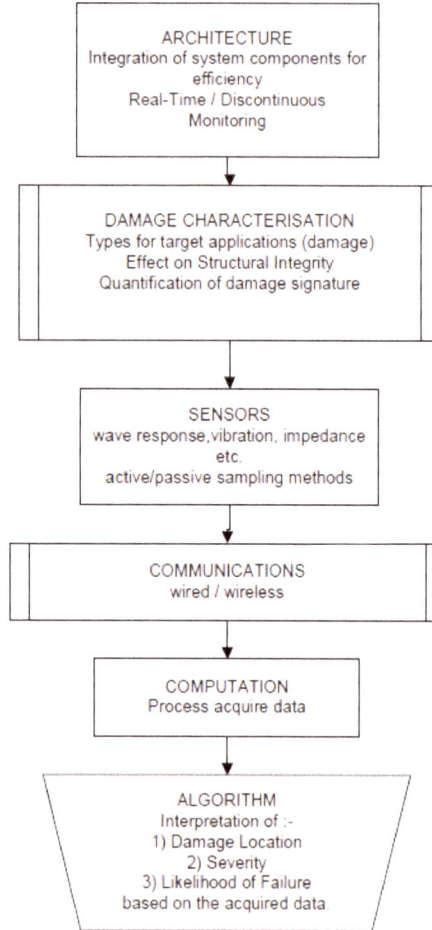

Figure 4: SHM system components introduced by Kessler [6].

5. Operational evaluation

Since one of the main roles of the structural monitoring is to acquire measurements on a structure for a period of time; it is essential that, this periodic measurement should be complement with the ideal operation of the structure. The key role of this assessment is to set limitations on what will be monitored and how the monitoring will be accomplished [75]. According to Bement and Farrar [76] there are four main questions or evaluation stages here that need to be answered, and they are:

1. The definition of the damage in the system being studied.

2. The economic and life safety justification for performing the monitoring procedures.

3. The environmental and operational conditions under which the system is to be monitored.

4. The limitations of acquiring data in the operational environment.

Undeniably, the long-term benefits of monitoring this structure with its anticipated or known operational and environmental conditions will significantly extend the life span of the system components and structures. Moreover, a strategic and reasonable monitoring schedule can be produced. One may consult Worden and Barton [77] for a thorough definition of the operational evaluation issue.

6. Data acquisition

Advanced technology in communications has prompted SHM researchers in telemetry technology to transmit and receive the acquired data from the monitored structure via wireless means [75]. References [78, 79] demonstrate how a wireless guided wave on the aircraft wing could be done remotely, left–in-place for in-situ defect detection. Nevertheless, there are indeed only two means of acquiring/transporting/transmitting a data to its designated storage medium which can be made, which are wired and wireless technology. The types of sensor to be used, sensor placement, number of sensors to be used, and the hardware implementations must be resolved first, before the decision on the types of transmitting devices can be made. Obviously, this data acquisition process requires an integrated framework between hardware and software elements, in order to develop cost-effective SHM systems.

Fundamentally, the hardware will be able to transmit (generating excitation force) and acquire the response data (sensing and acquiring) from the monitored structure to the hardware processor or vice-versa. On the other hand, the software will facilitate the operation of the data stored in the designated medium to be accessed and processed by the hardware processor. If a systematic and strategic combination between software and hardware can be made, the procedure in analysing and diagnosing the current state of the acquired data on the monitored structure can be made reasonably fast and in an accurate mode. Based on these requirements, the data acquisition process is application specific and requires economic considerations as stressed by Sohn *et.al.* [26]. In their paper, the consideration of normalising the input data is also discussed thoroughly and detailed analysis of this section can be obtained via these references as well [72, 80, 75].

7. Feature extraction and data condensation

Generally, feature extraction terminology comes from the pattern recognition literature and in short means *distinguishing feature* [77]. The distinguishing feature in this case is the portion of the acquired data that is sensitive to the detected damage. In other words, the extraction process involves the task of magnifying the characteristics of various damage classes and suppressing the normal background [77]. Examples of such extraction processes can be referred to in this reference. Since the criteria for inferring two obvious structural conditions are based on the acquired and extracted data with various operational and environmental conditions [75], the volume of the acquired data will inevitably be of a huge capacity. Therefore, for an efficient and systematic SHM system the capability of reducing or condensing the high dimension of the acquired data is desirable.

According to Farrar [75] the need to condense the data is advantageous and necessary if comparisons of many data sets over the lifespan of the structure are to be predicted or the data acquired will be measured over an extended period of time and in various operational environments. Principal Component Analysis (PCA) and Sammon Mapping techniques can be seen as the recommended reduction algorithm in condensing the acquired data. PCA is a multivariate statistical tool [81]. The main objective for a feature extraction module is to identify the most critical zone or wave packet (if one is applying a wave propagation technique) that is sensitive to damage. In addition, this extraction phase should be able to discard or alleviate the data that does not differentiate between the true states of the structural conditions.

8. Statistical model development

The last of the critical points in the SHM system is the statistical model development. This is where the proposed algorithm is required to perform the decision analysis where the diagnosis of the acquired, extracted and condensed data can be made. Since Farrar et al [75] outlined the damage identification problem as a 'pattern recognition' paradigm.

Worden and Barton [77] discussed three types of algorithms that can be applied here and they are:

i) Novelty detection

ii) Classification

iii) Regression

The novelty detection algorithm is the establishment of data when the undamaged or normal condition modes are only available for analysis. This is a two-class problem where unsupervised learning can be used. The classification and regression algorithm falls into supervised learning, where the damaged and the undamaged data are available for damage identification analysis. The most popular methods for the unsupervised learning are Outlier Analysis [82] and Probability Density Estimation Methods [83], whereas neural networks are the most common tools utilized in supervised learning [83, 84].

Based on the summary conducted previously, the four critical SHM system processes can be best described diagrammatically in a flowchart format as shown in Figure 5.

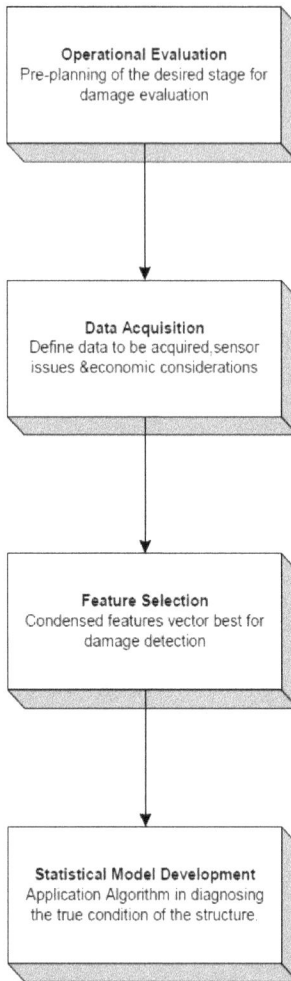

Figure 5 : SHM Pattern Recognition process variables

References

[1]   GmbH, E.D., The research requirements of the transport sectors to facilitate an increase usage of composite materials., in COMPOSITN. 2004.

[2]   Boller, C., (2001).Ways and options for aircraft structural health
      management.Smart Material Structure, 10: p. 432-440.

https://doi.org/10.1088/0964-1726/10/3/302

[3]   Fielding, J.P., Introduction to Aircraft Design. Vol. 1. 1999: Cambridge University
      Press.

https://doi.org/10.1017/CBO9780511808906

[4]   Boller, C., (2000).Next Generation structural health monitoring and its integration
      into aircraft design. International Journal of Systems Science, 31(11): p. 1333-
      1349.

https://doi.org/10.1080/00207720050197730

[5]   Sierakowski, R.L. and G.M. Newaz, Damage Tolerance in advanced composite.
      1995, Pennsylvania, USA: Technomic Publishing Company.

[6]   Kessler, S.S., PhD Thesis; Piezoelectric-Based In-Situ Damage Detection of
      Composite Materials for Structural Health Monitoring Systems, in Department of
      Aeronautics and Astronautics Massachusetts Institute of Technology. 2002,
      Massachusetts Institute of Technology: Massachusetts.

[7]   Raymer, D.P., Aircraft Design: A conceptual approach. 1992, Washington D.C:
      AIAA eductaion series.

[8]   Tober, G. and W.B. Klemmt. NDI Reliability Rules used by Transport Aircraft -
      European View Point. in 15th World Conference on Non-Destructive
      Testing.Rome.

[9]   Schmidt, H.J., B. Schmidt-Brandecker, and G. Tober, (1999).Design of Modern
      Aircraft Structure and the Role of NDI. NDT.net, 4(6).

[10] O' Brien, E., Structural Health Monitoring (Post-Nucleation Fatigue
      DamageDetection and Monitoring of Structures Designed by Damage Tolerant
      Principles). 2005, Airbus.

[11] Diamanti, K., J.M. Hodgkinson, and C. Soutis, (2004).Detection of Low-velocity
      Impact Damage in Composite Plates using Lamb waves. Structural Health
      Monitoring, 3(1): p. 33-41.

https://doi.org/10.1177/1475921704041869

[12] Bar-Cohen, Y., (1999).In-Service NDE of Aerospace Structures – Emerging
      Technologies and Challenges at the End of the 2nd Millennium. NDT.net, 4(9).

[13]   Tr`etout, H., (1998).Review of advanced ultrasonic techniques for aerospace structures. NDT.net, 3(9).

[14]   Website, http://www.ntsb.gov/publictn/publictn.htm accessed on 28/8/2005.

[15]   Staszewski, W.J., C. Boller, and G. Tomlinson, Health Monitoring of Aerspace Structures, Smart Sensor Technologies and Signal Processing. 2003: John Wiley.

[16]   Website, A., http://iac.dtic.mil/ntiac/aastory.htm accessed on 27/08/2005.

[17]   Chopra, I., (2002).Review of State of Art of Smart Structures and Integrated Systems. AIAA, 40(11): p. 2145-2187.

[18]   Mackerle, J., (2003).Smart materials and structures—a finite element approach— an addendum: a bibliography (1997–2002). Modelling and Simulation In Materials Science And Engineering, 11: p. 707-744.

https://doi.org/10.1088/0965-0393/11/5/302

[19]   Akhras, G., (2000).Smart Materials and smart systems for the future. Canadian Military Journal, (Autumn 2000).

[20]   Hall, S.R. and T.J.Conquest, (1999).The Total Data Integrity Initiative –Structural Health Monitoring, The Next Generation. Proceedings of the USAIF ASIP Conference.

[21]   Kessler, S.S. and S.M. Spearing, (2002).Damage Detection in Composite Material Using Lamb Wave Methods. Smart Materials & Structures, 11(2): p. 269-278. 209

https://doi.org/10.1088/0964-1726/11/2/310

[22]   Giurgiutiu, V. and J. Bao, (2004).Embedded-ultrasonic Structural Radar for In Situ Structural Health Monitoring of Thin-wall structures. Structural Health Monitoring, 3(2): p. 121-140.

https://doi.org/10.1177/1475921704042697

[23]   Van der Auweraer, H. and B. Peeters, (2003).International Research Projects on Structural Health Monitoring: An Overview. Structural Health Monitoring., 2(4): p. 341-358.

https://doi.org/10.1177/147592103039836

[24]   Balageas, D., (2002).Structural health monitoring R&D at the "European Research Establishment in Aeronautics"(EREA). Aerospace Science and Technology, 6: p. 159-170.

https://doi.org/10.1016/S1270-9638(01)01140-3

[25]   Ghosh, A. and Sinha P.K., (2004).Dynamic and impact response of damage laminated composite plates. Aircraft Engineering and Aerospace Technology, 76(1): p. 29-37.

https://doi.org/10.1108/00022660410514982

[26]   Sohn, H., et al., (2003).A Review of Structural Health Monitoring Literature: 1996-2001. Los Alamos National Laboratory Report, LA-13976-MS, 2003.

[27]   Staszewski, W.J., (2000).Monitoring on-line integrated technologies for operational reliability - MONITOR. Air & Space Europe, 2(4): p. 67-72.

https://doi.org/10.1016/S1290-0958(01)80019-8

[28]   Staszewski, W.J. Ultrasonic/Guided Waves for Structural Health Monitoring. In DAMAS. 2005. Gdansk, Poland.

[29]   Alleyne, D.N., PhD thesis, The Nondestructive Testing of plates using ultrasonic lamb waves, in Mechanical Engineering, Imperial College of Science, Technology and Medicine. 1991, University of London: London.

[30]   Website, N., http://www.asnt.org/ndt/primer4.htm accessed on 12/02/2004.

[31]   Li, J. and J.L. Rose, (2002).Angular-Profile Tuning of Guided Waves in Hollow Cyclinders Using a Circumferential Phased Array. IEEE Transcations of Ultrasonics, Ferroelectrics, and Frequency Control, 49(12): p. 1720-1724.

https://doi.org/10.1109/TUFFC.2002.1159849

[32]   Worden, K., (2001).Rayleigh and Lamb Waves -Basic Principles. Strain, 37(4).

https://doi.org/10.1111/j.1475-1305.2001.tb01254.x

[33]   Rose, J.L., Ultrasonic Waves in Solid Media. 1999: Cambrige University Press.

[34]   Krautkramer, Ultrasonic Testing of Materials. 2 ed. 1990: Springer Verlag.

[35]   Demer, L.J. and L.H. Fentnor, (1969). Lamb wave techniques in Nondestructive testing. International Journal of Nondestructive Testing, 1: p. 251-283.

[36]   Viktorov, I.A., Rayleigh and Lamb waves: physical theory and applications. Ultrasonic technology. 1967, New York: Plenum Press. x, 154.

https://doi.org/10.1007/978-1-4899-5681-1

[37]   Alleyne, D.N. and P. Cawley, (1992).The interaction of Lamb waves with defects. IEEE Transactions on Ultrasonics, Ferroelectrics, and Frequency Control: p. 381-397.

https://doi.org/10.1109/58.143172

[38]　Worden, K., et al., (2000).Detection of defects in composite plates using Lamb waves and novelty detection. International Journal of Systems Science, 31(11): p. 1397-1409.

https://doi.org/10.1080/00207720050197785

[39]　Staszewski, W.J. and B.C. Lee, (2002).Modelling of acousto-ultrasonic wave interaction with defects in metallic structures. ISMA2002.

[40]　Pierce, S.G., Culshaw B., Manson G., and W.K.a.W.J. Staszewski. The Application of Ultrasonic Lamb Wave Techniques to the Evaluation of Advanced Composite Structures. in SPIE International Symposium on Smart Structures And Materials 2000: Sensory Phenomena And Measurement Instrumentation For Smart Structures And Materials. 2000. Newport Beach, California, USA.

[41]　Kehlenbach, M. and D. S., (2002).Identifying Damage in Plates by Analysing Lamb Wave Propagation Characteristics. Proceedings of SPIE, Smart Nondestructive Evaluation for Health Monitoring of Structural and Biological Systems, 4702: p. 364-375.

https://doi.org/10.1117/12.469897

[42]　Liu, T., M. Veidt, and S. Kitipornchai, (2002).Single mode Lamb waves in composite laminated plates generated by piezoelectric transducers. Composite Structure, 58(3): p. 381-396. 210

https://doi.org/10.1016/S0263-8223(02)00191-5

[43]　Varadan, V.K. and V.V. Varadan. Wireless remotely readable and programmable microsensor and MEMS for health monitoring of aircraft structures. in International workshop on structural health monitoring, 2nd Conference. 1999.

[44]　ANSI/IEEE, An American National Standard IEEE Standard on Piezoelectricity ANSI/IEEE 176-1987. 1987, USA.

[45]　Worlton, D.C., (1961).Experimental Confirmation of Lamb Waves at Megacycle Frequencies. Journal of Applied Physics, 32(6): p. 967-971.

https://doi.org/10.1063/1.1736196

[46]　Alleyne, D.N.a.C.P., (1992).Optimization of Lamb wave inspection techniques. NDT & E international, 25(1): p. 11-22.

https://doi.org/10.1016/0963-8695(92)90003-Y

[47] Cawley, P., Alleyne, David N., (1996).The use of Lamb waves for the long range inspection of large structures. Ultrasonics, 34: p. 287-290.

https://doi.org/10.1016/0041-624X(96)00024-8

[48] Nishino, H., T. S., F. Uchida, M. Takemoto, and K. Ono, (2001).Modal Analysis of Hollow Cylindrical Guided Waves and Applications. The Japan Journal Applied Physics, 40(1): p. 364-370.

https://doi.org/10.1143/JJAP.40.364

[49] Seale, M.D., B.T. Smith, B.T. Processer, and J.E. Masters, (1994).Lamb Wave Response of Fatigued Composite Sample. Review of Progress in Quantitative Nondestructive Evaluation., 13B: p. 1261-1266.

[50] Pei, J., et al., (1995).Lamb wave tomography and its application in pipe erosion/corrosion monitoring. IEEE Ultrasonics Symposium, 6: p. 795-798.

https://doi.org/10.1109/ultsym.1995.495686

[51] Lee, Y.C. and S.W. Cheng, (2001).Measuring Lamb Wave Dispersion Curves of a Bi-Layered Plate and Its Application on Material Characterization of Coating. IEEE Transcations of Ultrasonics, Ferroelectrics, and Frequency Control, 48(3).

[52] Na, W.-B., T. Kundu, and M.R. Ehsani, (2003).Lamb waves for detecting delamination between steel bars and concrete. Computer-Aided Civil and Infrastructure Engineering, 18(1): p. 58-63.

https://doi.org/10.1111/1467-8667.t01-1-00299

[53] Hansch, M.K., K.M. Rajana, and J.L. Rose, (1994).Characterization of Aircraft Joints using Ultrasonic Guided Waves and Physically based Feature Extraction. IEEE Ultrasonics Symposium: p. 1193-1196.

https://doi.org/10.1109/ultsym.1994.401799

[54] Alleyne, D.N. and P. Cawley, (1990).A 2-Dimensional Fourier Transform method for the quantitative measurement of Lamb modes. IEEE Ultrasonics Symposium: p. 1143-1146.

https://doi.org/10.1109/ULTSYM.1990.171541

[55] Wilcox, P.D., M.J.S. Lowe, and P. Cawley, (2001).Mode and Transducer Selection for Long Range Lamb Wave Inspection. Journal of Intelligent Material Systems and Structures., 12: p. 553-565.

https://doi.org/10.1177/10453890122145348

[56]   Worden, K., D. Allen, H. Sohn, and C.R. Farrar, Damage Detection in Mechanical Structures using Extreme Values Statistics. 2002: Los Alamos National Laboratory report LA-13903-MS.

[57]   Finlayson, R.D., M. Friesel, M. Carlos, P. Cole, and J.C. Lenain, (2001).Health monitoring of aerospace structures with acoustic emission and acousto-ultrasonics. Insight, 43(3).

[58]   Geng, R.S. Application of Acoustic Emission For Aviation Industry- Problems and Approaches. in World Conference on NDT. 2004. Montreal Canada.

[59]   Manson, G., K. Worden, A. Martin, and D.L. Tunnicliffe, (1999).Visualisation and dimension reduction of acoustic emission data for damage detection. Key Engineering Materials, 167-168: p. 64-75.

https://doi.org/10.4028/www.scientific.net/KEM.167-168.64

[60]   Tscheliesnig, P. Corrosion Testing of Ship Building Materials with Acoustic Emission. in 26th European Conference on Acoustic Emission Testing. 2004. Berlin. 211

[61]   Website, A., http://www.capgo.com/Resources/ConditionMonitoring/Acoustic.html accessed on 23/10/2005.

[62]   Website, E., http://www.engineersedge.com/inspection/eddycurrent.htm accessed on 12/11/2005.

[63]   Schall, W.E., Non-Destructive Testing, ed. M.P. Co.Ltd. 1968, London.

[64]   Website, E., http://www.ndt.net/article/wcndt2004/eddy_current.htm accessed on 12/11/2005.

[65]   Titman, D.J., (2001).Applications of thermography in non-destructive testing of structures. NDT&E International, 34: p. 149-154.

https://doi.org/10.1016/S0963-8695(00)00039-6

[66]   Website, V.T., http://www.asnt.org/publications/materialseval/basics/jul98basics/jul98basics.html accessed on 15/11/2005.

[67]   Mufti, A.A., (2002).Structural Health Monitoring of Innovative Canadian Civil Engineering Structures. Structural Health Monitoring, 1(1): p. 89-103.

https://doi.org/10.1177/147592170200100106

[68]   Keller, E. and A. Ray, (2003).Real-time Health Monitoring of Mechanical
       Structures. Structural Health Monitoring, 2(3): p. 191-203.

https://doi.org/10.1177/1475921703036048

[69]   Schulz, M.J., A. Ghoshal, M.J. Sundaresan, P.F. Pai, and J.H. Chung,
       (2003).Theory of Damage Detection Using Constrained Vibration Deflection
       Shapes. Structural Health Monitoring, 2(1): p. 75-99.

https://doi.org/10.1177/147592103031114

[70]   Carden, E.P. and P. Fanning, (2004).Vibration Based Condition Monitoring: A
       Review. Structural Health Monitoring, 3(4): p. 355-377.

https://doi.org/10.1177/1475921704047500

[71]   Ching, J. and L.J. Beck, (2004).New Bayesian Model Updating Algorithm Applied
       to a Structural Health Monitoring Benchmark. Structural Health Monitoring, 3(4):
       p. 313-332.

https://doi.org/10.1177/1475921704047499

[72]   Doebling, S.W., C.R. Farrar, and M.B. Prime, (1996).A summary review of
       vibration based damage identification methods. The Shock and Vibration Digest,
       30(2): p. 91-105.

https://doi.org/10.1177/058310249803000201

[73]   Mal, A., F. Ricci, S. Banerjee, and F. Shih, (2005).A Conceptual Structural Health
       Monitoring System based on Vibration and Wave Propagation. Structural Health
       Monitoring, 4(3): p. 283-294.

https://doi.org/10.1177/1475921705055254

[74]   Farrar, C.R. and H. Sohn. Pattern Recognition for Structural Health Monitoring. in
       Workshop on Mitigation of Earthquake Disaster by Advanced Technologies. 2000.
       Las Vegas, USA.

[75]   Farrar, C.R., T.A. Duffey, S.W. Doebling, and D.A. Nix. A Statistical Pattern
       Recognition Paradigm for Vibration-Based Structural Health Monitoring. In
       Proceedings of the 2nd International Workshop on Structural Health Monitoring.
       2000. Stanford, CA, USA.

[76]   Bement, M.T. and C.R. Farrar. Issues for the Application of Statistical Models in
       damage detection. in International Modal Analysis Conference (IMAC 18). 2000.
       San Antonio.

[77] Worden, K. and J.M. Dulieu-Barton, (2004).An Overview of Intelligent Fault Detection in Systems and Structures. Structural Health Monitoring, 3(1): p. 85-98. https://doi.org/10.1177/1475921704041866

[78] Sazonov, E., K.D. Janoyan, and R. Jha. Wireless Intelligent Sensor Network for Autonomous Structural Health Monitoring. in Smart Structures/NDE 2004. 2004. San Diego, California.

[79] Zhao, X., C. Kwan, and M. Luk. Wireless Nondestructive Inspection of Aircraft wing with ultrasonic guided waves. in 16th World Conference on NDT. 2004. Montreal,Canada.

[80] Farrar, C.R. and S.W. Doebling. An Overview of Model-Based Damage Identification Methods. in DAMAS. 1997. Sheffield, UK. 212

[81] Jolliffe, I.T., Principal Component Analysis. 1986: Springer-Verlag. https://doi.org/10.1007/978-1-4757-1904-8

[82] Worden, K., G. Manson, and N.R.J. Fieller, (2000).Damage detection using outlier analysis. Journal of Sound and Vibration, 229(3): p. 647-667. https://doi.org/10.1006/jsvi.1999.2514

[83] Bishop, C.M., (1994).Novelty detection and neural network validation. IEEE Proc. Vision and Image Signal Processing, 141: p. 217-222. https://doi.org/10.1049/ip-vis:19941330

[84] Sohn, H., C.R. Farrar, N.F. Hunter, and K. Worden, (2001).Structural Health Monitoring using statistical pattern recognition techniques. Journal of Dynamics System Measurement and Control, 123: p. 706-711. https://doi.org/10.1115/1.1410933

[85] Kessler, S.S. and D.J. Shim. Validation of a Lamb Wave-Based Structural Health Monitoring System for Aircraft Application. in SPIE Conference. 2005. https://doi.org/10.1117/12.599757

[86] Manson, G., K. Worden, and D. Allman, (2003).Experimental validation of a structural health monitoring methodology. Part I. Novelty detection on a Gnat aircraft. Journal of Sound and Vibration, 258(2): p. 345-363. https://doi.org/10.1006/jsvi.2002.5167

[87] Giurgiutiu, V., A. Zagrai, and J. Bao, (2004).Damage Identification in Aging
      Aircraft Structures with Piezoelectric Wafer Active Sensors. Journal of Intelligent
      Material Systems and Structures, 15: p. 673-687.

https://doi.org/10.1177/1045389X04038051

[88] Yang, J., F.K. Chang, and M. Derriso, (2003).Design of a Hierarchical Health
      Monitoring System for Detection of Multilevel Damage in Bolted Thermal
      Protection Panels: A Preliminary Study. Structural Health Monitoring., 2(2): p.
      115-122.

https://doi.org/10.1177/1475921703002002003

[89] Ihn, J.B. and F.K. Chang, (2004).Detection and monitoring of hidden fatigue crack
      growth using a built-in piezoelectric sensor/actuator network: I.Diagnostics. Smart
      Material Structure, 13: p. 609-620.

https://doi.org/10.1088/0964-1726/13/3/020

[90] Ihn, J.B. and F.K. Chang, (2004).Detection and monitoring of hidden fatigue crack
      growth using a built-in monitoring piezoelectric sensor/actuator network:
      II.Validation using riveted joints and repair patches. Smart Material Structure, 13:
      p. 621-630

https://doi.org/10.1088/0964-1726/13/3/020

Chapter 2

An overview of structural health monitoring: from hard time to online monitoring

[1]K. D. Mohd Aris, [2]A.Hamdan, [2]F. Mustapha

[1]Unversiti Kuala Lumpur Malaysian Institute of Aviation Technology, 2891, Jalan Jenderam Hulu,43800 Dengkil, SelangorbMalaysa

[2]Department of Aerospace Engineering, Universiti Putra Malaysia, 43400 Serdang, Selangor, Malaysia,

Keywords

Structural Health Monitoring (SHM), Non-Destructive Inspection, Condition Monitoring, Composite Material

Abstract

Structural Health Monitoring (SHM) is defined as the "acquisition, validation and analysis of technical data to facilitate life cycle management decisions". In addition, SHM denotes as a system with the ability to detect and interpret adverse "changes" in a structure in order to improve reliability and reduce Life Cycle Costs. The most fundamental challenge in designing an SHM system is knowing what "changes" to look for and how to identify them. The characteristics of damage in particular structures play a key role in defining the architecture of the SHM system. The resulting "changes," or damage signature, will dictate the type of sensors that are required, which determines the requirement for the rest of the components in the systems. Next, the scope of condition based monitoring (CBM) is discussed through the use of the various non-destructive inspection (NDI) techniques which are available for damage detection on advanced composite structures both at present and in the near future. In addition, the limitation of current NDI techniques is discussed and how this has created the path for the Structural Health Monitoring to be implemented.

Contents

1.   Non-destructive inspection - present damage detection

Throughout the years, NDI has been the most important tool used in order to assess the condition of the aircraft structure. NDI inspections are being accepted as part of procedures during major aircraft inspection tasks such as the C-checks, Heavy Maintenance Visit (HMV) or during defect findings, as has been emphasized by Grandt Jr. (2011). Rectification of the damage has to be accompanied by task completion documentations when the Certificate of Release to Service (CRS) is issued as stated in the Laws Of Malaysia - Act 3: Civil Aviation Act 1969, The Commissioner Of Law Revision, 2006. The improvement in NDI technology has allowed inspection to be tailored to the materials used, the operational conditions and the level of difficulties. However, an autonomous NDI is privileged to large aircraft manufacturing or assembly due to its high cost in obtaining certification, implementing the system and training for competent human capital. At an operational level, the NDI application is still dependent on human capital skills and is carried out while the aircraft is on the ground. This is achieved by detecting, locating and sizing any detected flaws. (Halmshaw, 1991) There are several inherent difficulties in detecting damage in composite materials as opposed to traditional engineering materials such as metallic or plastics. One reason for this is the inhomogeneity and anisotropy of composites; most metals and plastics are formed by one type of uniformly isotropic material with very well-known properties. Laminated composite materials, on the other hand, can have a wide varying set of material properties based on the chosen fibers, matrix and manufacturing process. This makes modelling for composites more complex, often involving non-linear and hybridization of materials such as fiber, matrix, core, thixotropic agents etc. In addition, any damage occurring in composite materials can be on the surface, embedded or on the adjacent side. The damage that cannot be seen directly and is hidden is known as Barely Visible Internal Damage (BVID). This type of damage can prevent or limit the implementation of several detection methods (Scott and Scala, 1982). The importance of damage detection for composite structures is often accentuated over metallic or plastic structures because of their load bearing requirements. Typically, unreinforced plastics are not used in load critical members; since their properties are predictable and they are usually simple and inexpensive to manufacture, they are often designed to be replaceable, safe- life parts. Similarly, metals are generally well understood and easy to model, thus they are frequently designed using damage tolerant methodologies. The behavior of composite material, on the other hand, is much less well understood, and an unexpected failure of the composite part could prove catastrophic to a vehicle. New techniques are emerging, which augment the current techniques, in order to ensure that the probability of damage (POD) within the structure (such as delamination, disbonding and voids) is more effective for modern composite structures, e.g. wind generation turbine blades, aircraft pressurized fuselages and civil structures. Studies have been carried out by Amenabar et.al. (2011) and Mahoon (1998) in detecting damages on the above structure. Therefore, the development of a reliable damage detection method is critical to maintain the integrity of aerospace vehicles as revealed by Heida and Platenkamp (2011). The following sections provide descriptions of various non-destructive techniques that have been developed for the detection of damage in composite materials.

2. Visual inspection method

Perhaps the most natural form of evaluating composite structures is by visual inspection (Dorworth et. al, 2009). There are several variants of this method existing at various levels of sophistication, from the use of a static optical or scanning electron microscope to optical examination by eye over the structure. While microscopy can be a useful method to obtain detailed information such as micro-crack counting or the delamination area, it can only be used in the laboratory therefore requiring that any given section must be removed from the larger structure. Visual inspections on particular structures are perhaps the simplest and least expensive method; they are relatively fast and capable of detecting relevant impact damages (Non-destructive inspection, 2005). However, any damages which have occurred in the sub-laminate layer or on the other side of the surface are difficult to detect by eye and may require additional equipment such as a flash light, a magnifying glass, or borescope etc. In addition, visually, the eyes alone can only determine very little detail about the damage mechanism or its severity. While this method can potentially provide some useful data for damage detection, on a large-scale structure this process would prove inefficient and ineffective as determined by Campbell (2003, pp. 471-512).

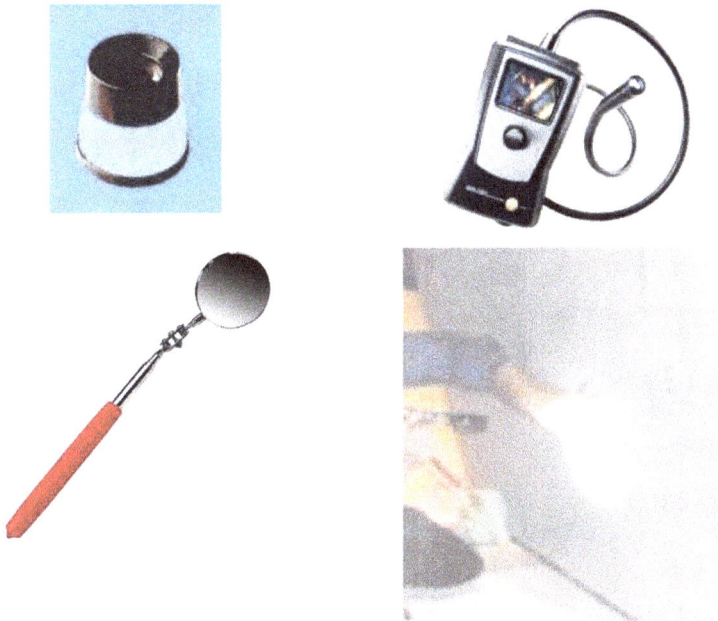

Figure 1 Typical visual inspection carried out on composite structures by: a. magnifying glass, b. borescope/videoscope, c. inspection mirror and d. light assistance. (Dorworth et. al, 2009)

3.   Tap test methods

This is the most well-known, vibration-based method specifically used on composite structures. The inspection uses a small metallic shape like a coin which is then lightly tapped on the suspected structure. The change in sound between defective and defect-free regions indicates the presence of damage. However, the sensitivity of the method decreases with a defect depth of 1mm under the skin or the sub laminate area as found by Cawley and Adams (1989). In addition, the inspection only provides a rough dimension of the damaged section and is limited by approved certified personnel. Advanced equipment that digitally records the thudding sound and displays it has replaced the coin tap test. The digital tap test captures the tapping sound made by the tap hammer connected to a unit which translates the tapping energy to sound energy. The sound energy is displayed as a numerical value. During inspection, the tap hammer is tapped at the adjacent area or on a reference specimen for a baseline reading. Georgeson, Lea and Hemsen (2013) pointed out that a large variation between the reference area and the suspected area is an indication of delamination.

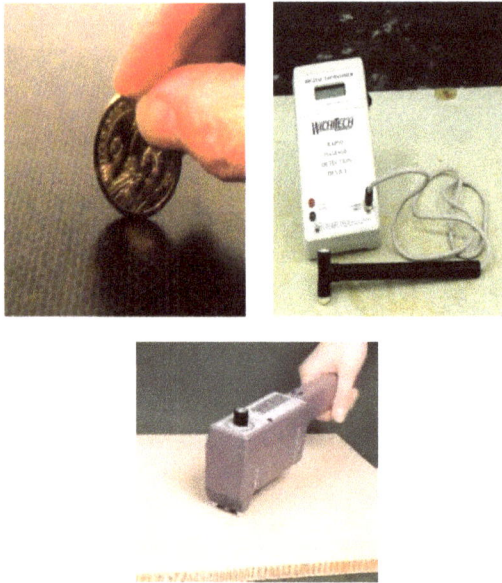

Figure 2 Typical tap test equipment available for the aviation field. a. Coin tap test, b. Digital tap hammer from Wichitech, and c. Woodpecker from Mitsui.

An automated tap test can be conducted by using a Woodpecker WP632 lightweight, hand-held device that uses a battery driven solenoid hammer with a force sensor built in the hammer tip (http://wp632.cadex.co.jp/products/wp632am.htm). It operates by measuring the hammer sound reflection time once the tapping is initiated by the vibrating plunger in contact with the surface of the test part. The time of the tapping sound increases with respect to the existing defects such as disbonding or delamination.

However, tap testers are entry level inspection items and they are best used to detect damage existence, without having the ability to locate and map the damage area. The detectability efficiency for disbonding and delamination is varied and not always consistent.

4.   X-ray inspectionethods

X-ray technique relies on recording the difference in x-ray absorption rates through the surface of a structure. These methods are implemented either in real-time digitally, or by taking static radiographs, where areas of different permeability or density are differentiated by the magnitude of x-ray exposure to the media on the opposite side of the surface after a predetermined excitation time. To accentuate damaged regions with cracks or delamination, a liquid penetrant is often applied to the area to be examined (Dance, 1976). While these techniques are relatively inexpensive and simple to implement and interpret, they require large and costly equipment that is difficult to use on large structural components without removing them from the vehicle. Other limitations include their inability to detect defects in thin laminate structures, orientation variations, hybrid materials and image quality indicator (Jones and Polansky, 1988). The greatest challenge to using x-ray in a vehicle inspection application is that all of these methods require access to both sides of the surface in order to emit and collect the X- ray radiation, which is often not practical. In addition, safety and hazard issues affect the operation time of the procedure, which has to be performed inside a closed hangar with only a handful of NDI inspectors performing the task.

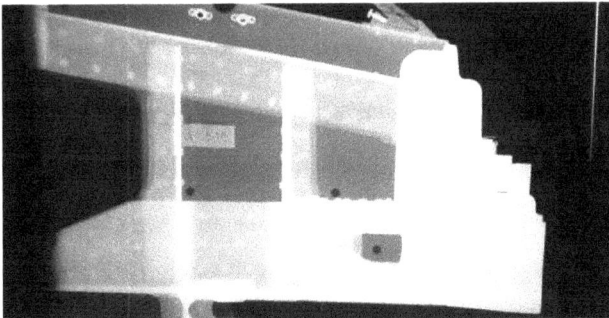

Figure 3 Typical x-ray inspection result on an aircraft component (Downloaded from http://ndtaviation.com/rt-x-ray-aviation).

5.   Ultrasonic methods

Another commonly implemented NDI technique is ultrasonic testing, most often referred to as A-, B- and C-scans. The method uses an ultrasonic signal and measures the attenuation of the signals using stress waves on the inspected structure (Růžek, Lohonka, and Jironč, 2006). The stress waves are mechanical waves or vibrations in which for composites the compatible frequency range is between 1 to 10 MHz. There are two modes of operation which are pulse echo and through transmission. A pulse echo uses a single transducer which behaves as a transmitter and receiver. Alternatively, through transmission uses two transducers which are placed in areas across the inspected panel. The transmission of the ultrasonic waves is achieved by the use of a couplant such as liquid and water but these couplant may contaminate the structure further if the couplant is seeping through the unseen damage. Newer techniques, such as non-contact couplant, uses air to transmit the waves, but more confident results are required for the method to be used on the current NDI for aircrafts. An A-scan refers to a single point measurement of density, a B-scan measures these variations along a single line, and a C-scan is a collection of B-scans forming a surface contour plot. The C-scan has been common practice in the aerospace industry since the introduction of composite parts to this field, since its results are widely understood and can be used to scan a large area of structure in a relatively short time period with the ability for defect detection, sizing and depth (Hsu, 2008).

Although the use of ultrasonic inspection has more accurate results compared to the other method discussed above, but are limited to laminated structure only. Further works by Ali et. al. (2012) found that any disbonding existing between the skin and the honeycomb will have a return signal which is larger than the bonded skin and honeycomb because the signal returns from the upper skin's lower surface without being transmitted down and up the honeycomb wall. Another setbacks, is that the method requires different calibration standards and recalibration based on each configuration's materials and thickness as supported by the work of Ren and Lissenden (2013), Raišutis et. al. (2010) and Bar-Cohen and Crane (1985).

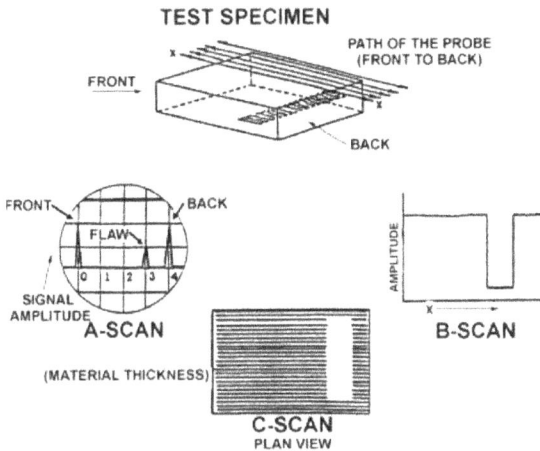

Figure 4 Typical scanning method for ultrasonic inspection (Downloaded from http://www.engineersedge.com/inspection/ultrasonic.htm).

6. Other types of NDT

Technological breakthroughs in NDI methods are evolving due to the challenges posed in applying the composite structure. Challenges such as thicker structural materials, the hybridization of different types of material configurations, and the area of the materials utilization require new types of inspection approaches and techniques. Thermography (Maldaque et.al, 2004, Bendada et.al, 2013 and Usamentiaga et. al., 2012), Phased array (Holmes et. al., 2008, Drinkwater and Wilcox, 2006, and Duxbury et. al., 2013, laser shearography (Junyan et. al., 2013 and Genest et.al, 2009 and acousto imaging (Lasser et. al., 2010, King et. al., 2003, are some of techniques still in their development phase and are evolving in order to support the aircraft maintenance field.

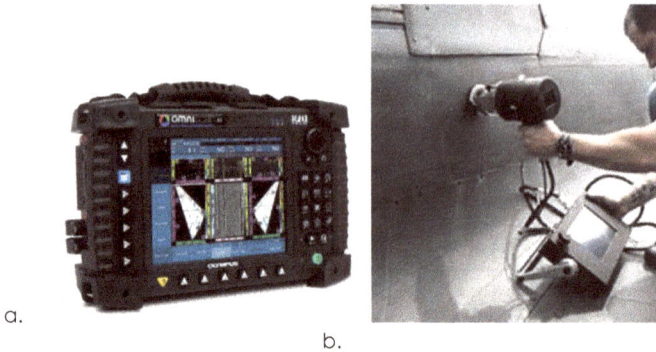

Figure 5 An example of a. Phase array inspection (http://www.olympus-ims.com/en/ omniscan-mx2/), and b. Acousto imaging (http://www.imperiuminc.com/).

7. Damage detection techniques

The main objective of damage detection techniques is to detect any anomalies from the baseline pattern, which are the indications of damage presence as pointed out by Jardine, et. al. (2006). The degree of damage will have an immediate or prolonged effect on a structure. Damage identification is carried out in conjunction with five closely related disciplines that include Structural Health Monitoring (SHM), Condition Monitoring (CM), Non-Destructive Evaluation (NDE), Statistical Process Control (SPC), and Damage Prognosis (DP), as concluded by Farrar et. al. (2009). Typically, SHM is associated with on-line, autonomous global damage identification in structural systems. CM is analogous to SHM, but addresses damage identification in rotating and reciprocating machinery. NDE is usually carried out off-line in a local manner with some a priori knowledge of the damage location. SPC is process-based rather than structure-based and uses a variety of sensors to monitor changes in a process, one cause of which can result from structural damage. However, many of the statistical monitoring tools developed for SPC have been adapted to SHM and CM applications.

When damage can be detected at its earliest possible stage and an immediate response takes place, it can significantly reduce the maintenance cost. Previous maintenance philosophy is based on hard time (HT) in order to prevent any catastrophic events. Parts are removed and replaced within the stipulated time or after certain usage cycles. The schedule is based on the theoretical life retrieved from a fatigue test with an added safety factor (McDonald, 2001; and Ghobbar and Friend, 2002). This leads to good parts being removed and this contributes to the high cost of operation. In order to save cost, selected parts, especially the rotating machinery, were put under condition monitoring (Jonathan and Burrows, 1994; Elforjani et. al, 2012). The parts were allowed to operate until anomalies were detected. In terms of structure, the components were put on NDI after a certain interval or any damage emerged, as studied by Wu and Syau (1993).

However, these two systems require the part or component to be shut down for inspection. This is known as down time and delays operation, which leads to loss of profit. The longer the downtime, the more revenue is lost. The new philosophy is to actively monitor the component, which is known as Structural Health Monitoring or SHM (Worden and Manson, 2007). This system allows monitoring in offline, online, active and passive modes. Figure 6 shows the evolution of damage detection systems.

Figure 6 The evolution of damage detection techniques (Worden and Manson, 2007).

8. Conditioning monitoring

Condition monitoring was initially used to detect damage by monitoring the rotating machine's performance through the vibration that it produces. Condition monitoring is described as a means to prevent the catastrophic failure of critical rotating machinery using a maintenance scheduling tool that uses vibration, infrared or lubricating oil analysis data to determine the need for corrective maintenance actions as pointed out by Adam et. al. (1978) and Davis (1998). The concept of CM dates back to the 1960s when component vibrations were monitored for US Navy ships, petrochemical industries and power generations companies. They invested a large amount of research and development into the development of analysis techniques based on vibration or noise that can be used to detect incipient problems in the critical mechanical components. By the 1980s a strong foothold in equipment and analytical skills was fully developed to analyse vibration analysis.

However, due to the high cost in implementing this method, only selective critical equipment is equipped with the system as stated by Becker and Rauber (2010). CM has successfully been used in monitoring rotating component using pattern recognition techniques. Any component with a diverging pattern will be isolated, rectified and repaired as examined by Elangovan et. al. (2011).

9.   Non destructive inspection (NDI)

The history of NDI can be traced back to the middle of the nineteenth century when a boiler at the Fales and Gray Car Inc. exploded and caused many casualties. A regulation was passed with a requirement for boilers to be inspected annually by visual inspection as stated by Breysse (2012). As a result of that incident, many techniques were developed and employed to support inspections of typical metallic structures. The use of NDI on composite structures can be traced back to the late 1960s when they were using immersing ultrasonic inspection on a fibrous composite structure (Hislop, 1969). The advantages of NDI techniques lie in their detectability, localization and sizing of the detected flaws but the utmost benefit is that the inspected area is free from damage (Halmshaw, 1991). Until the present, many NDI techniques have been improved or have emerged in various fields to fulfill the challenges of materials advancement and improvement in order to detect damage during production or maintenance. New materials such as composites are superior to metallic structures. This technology breakthrough has challenged the NDI field due to its compatibility issues, laminated plies, fiber orientation, imperfections within layers etc.; such damage is non-existent in metallic structures (Scott Scala, 1982). Although NDI technologies have nearly matured and more techniques have emerged, they still rely on human interference and the tasks are very localized when they are carried out. Unfortunately, all damages are hidden and can only be detected once they have occurred or have propagated. Currently an autonomous system is hardly likely to satisfy the inspection of the structure due to curvature, hidden places and equipment size. Therefore, automated and online active systems are needed to alert the aircrew and maintenance crews of the health status of the structure.

10.   Structural health monitoring (SHM)

Both of the damage detection techniques discussed above have been used for quite some time. However they are suitable for rotating machinery and structural integrity assessment respectively. According to Farrar and Worden (2007), SHM combines both damage detection philosophies and translates them into an active/passive of online/offline systems in order to monitor the health status of the structure as shown in Figure 7:

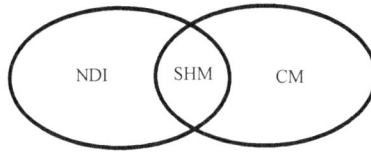

Figure 7 SHM philosophy (Farrar and Worden, 2007).

The advancement in sensor technology and data acquisition apparatus has made this approach feasible as the related sensors are becoming smaller and computing power is increasing respectively. Since the sensors are small, they can be embedded or on-surface installed without giving a tremendous weight penalty. The main aircraft manufacturers, such as Boeing (Benedettini, et. al, 2009) and Airbus (Speckmann, 2007), are gearing toward implementing this technology as the material structures become more advanced.

The use of composite materials has further demanded a suitable technique to detect any incipient damage within the structure or on the surface. Both companies have embarked on this technology on some of the current flight test aircrafts. The full implementation of the SHM systems will be seen in their newly developed aircraft (known as smart aircraft) within the next ten years when the long range single aisle replacement aircraft takes to the skies, for example the Boeing 787 and the Airbus A350 (Diamanti and Soutis, 2010).

However, most of the SHM damage detection systems, using various methods and sensor technologies, mainly assess the state of the parent structure. The health or condition is evaluated because the response behaves differently when any damage exists. During the operation of the aircraft, the external structures are exposed to the environmental conditions, whereas the internal structures are exposed to the system within the aircraft. Both types of damage, if undetected during the walk-around check or during the maintenance visit, may lead to catastrophic events. Earlier examples of this were when fatigue damages were not well understood, e.g. the DeHavilland Comet (Schijve, 1994), smokey rivets being undetected on Aloha Airlines B737 aircraft (Hendricks, 1991), Jack screw housing failure on the Alaska Airline (Dekker, 2004). From these examples, the need for effective damage detection is highlighted to ensure the safety of the passengers and crews during the flight.

There are challenges for any type of damage repair carried out to return the structure back to its original strength. Diamanti and Soutis (2010) define the challenges in applying SHM to an aero-structure as follows: the performance and integration of the newly repaired structure with the existing SHM system, new surface integrity due to the change in ply thickness and orientation, and the disrupted baseline model due to the repair, etc.

11. Structural health monitoring overviews

An aircraft use highly complex systems in which the redundant failure detection mechanism has been implemented in order to increase the capability of damage detection. These redundant systems are isolated from each other. The principle of this is that if one of the systems fails, the remaining system can still perform the vital task of ensuring that the aircraft can be flown and landed safely. Systems such as Engine Condition Monitoring (ECM) (Turner and Bajwa, 2008), Built in Test Equipment (BITE) (Johnson, 1996), Health Usage and Monitoring System (HUMS) (Wallace et. al., 2004) are already installed on some of the current flying fixed wing and rotary wing aircraft systems. Nevertheless, such damage detection on aircraft structures only exists on newly developed military aircrafts but not on civilian aircrafts. Furthermore, the technology involved in aircraft structure construction materials has changed drastically from utilizing wood and fabrics to aluminum alloys. Newly developed materials, such as Carbon fiber reinforced plastic (CFRP), Glass Aluminum Reinforcement (GLARE) and aluminum–lithium alloys, are starting to replace the traditional aluminum alloys. These changes are mainly to exploit their superiority of strength, stiffness, and tailorability, and to reduce the corrosion usually found in aluminum alloys. Both major aircraft manufacturers (Airbus and Boeing) are competitively introducing SHM technology into their next respective aircraft developments. Speckman (2007) indicated that Airbus has targeted that by the year 2018, an in service aircraft with fully integrated SHM sensor systems will be put on test flight, whereas according to Lopz and Klijn (2010) and Beneditti (2009), Boeing has taken a more global approach by compiling the monitoring system into the Integrated Vehicle Structural Health Monitoring System (IVSHM). Eventually, both manufacturers are aiming to implement the SHM systems on their next generation aircraft which will lead to more research and funding opportunities during the course of their implementation. As more new aircraft are delivered in the future, SHM will assist in terms of providing an efficient maintenance program. In addition, the current aircraft maintenance philosophy will be translated to condition monitoring as more technology is incorporated on the aircraft in order to provide the health status from the structure and system point of view.

In order to ensure that the aircraft are airworthy, they need to be maintained at regular intervals, the frequency of which depends on the manufacturer's recommendations and the experience of the aircraft operators. Sriram and Haghani (2003) stated that the best maintenance interval is very much related to the type of materials, the environmental conditions in which the aircraft is operated, the number of service cycles and the age of the aircraft. Alternatively, Mahoon (1998) identified that the current techniques to monitor the health of the aircraft structures are only by using visual and NDI techniques. The NDIs are carried out through air crew reports or maintenance schedules. The nature of the inspections is passive and the aircraft need to be grounded either during the hangar check visit or in service inspection. It requires specialized skills and expensive equipment to detect the damage in the suspected area. The use of advanced composite materials has introduced more challenges in detecting this damage. Surface damage due to impact can lead to hidden damage that cannot be visibly detected. Polimeno and Meo (2009) found that an indication of high impact can lead to a visible penetration on the structure, however a slow impact may lead to an internal matrix crack, known as barely visible impact damage (BVID). A thorough inspection has to be carried out on this type of damage in order to determine the extent of the true damage internally. The use of advanced NDI techniques, such as ultrasonic, radiography, thermography, and laser shearography, may assist in detecting this hidden damage but time constraints, skilled personnel and grounded aircraft are the drawbacks of the present approach (Adams and Cauley, 1988)

Online monitoring of the structure (known as SHM) is a new paradigm to ensure that the conditions of the aircraft structures are at the best operating parameters. Any presence of damage due to internal propagations or external impact can be detected through online or offline means in either passive or active modes. An ideal system can localize the damage, and also the degree of the damage can be systematically categorized. Ahmad et. al. (2000) and Stazewski et. al. (2009) have concluded that the initiation of the damage on the structure can be contained at the earliest time possible in order to save cost and manpower for the aircraft operator.

12. Motivation in using SHM

SHM is an emerging technique related to the development and application of detecting damage which is either autonomous or semi-autonomous. Three types of design approach are being used in aerospace structures as indicated by researchers and shown Figure 8. The safe life approach was the initial design methodology of the aircraft components (dated back to 1960). The safe life approach is based on the concept that any substantial defect which can occur on an aircraft, such as fatigue cracking, will not form during the service life of a component. The life is determined from fatigue test data (S–N curves) and the calculations are done using a cumulative damage "law". The design Safe-Life is then obtained by applying a safety factor (Ward and Parish, 1969). The determination of the serviceable life is done based on analytical and experimental analysis during the design stage. A Factor of Safety (FS) value between 2 to 4 is used to ensure a safe life operation before the structure or component fails. When the service life is equal to the design safe life the part has to be replaced accordingly to ensure that the predicted failure does not occur in the part. The advantage of this method is that it is a very simple model to design and it is conservative. It reduces the inspection time and cost due to its hard time component replacement. This method is suited to being applied to critical systems that are difficult to repair or can cause severe damage to life or property, according to Amer (1998). Nevertheless, this approach has the disadvantage that hard time replacement can lead to a good part being discarded as a result of its due time. Suresh (1998) and Nyman (1996) stated that it increases the operation cost because of the predictive time based approach of the safe life implication and the products' life are over-estimated. Another drawback is that damage sometimes occurs prematurely or unpredictably as determined by the Federal Aviation Administration (FA, 2012).

Figure 8 Damage detection approach (Gudmundsson, 2014; Price et. al., 2006, Curran, 2006, Dilger et. al., 2009 and Tan, Chen and Jin, 2005).

Currently, the safe life approach is too conservative; the Federal Aviation Administration (FAA) through its Federal Aviation Regulation (FAR) under FAR 25.57 stipulated that only landing gear and its attachments are required to be monitored by the safe life method. Greenbank (1991) and the Federal Aviation Administration (2012) showed that inspection of other components is based on the operator reliability study in which a new maintenance schedule has to be created and approved by the authority to ensure the safety of the aircraft is not compromised. This is due to different aircraft operators operating their aircraft based on their business model. A low cost flight may be subjected to a high flight cycle due to the route coverage and short flight duration, whereas for longer routes, the aircraft spends most of its time flying.

A damage tolerant design is the ability of the structure to sustain defects until further action is taken to bring the structure back to its operational status. The philosophy assumes that damage such as cracks or flaws can occur and develop during the service life of the component. A structure is considered to be damage tolerant if a maintenance program has been put in place in order to detect damages by routine inspection and repair any incidental damage, fatigue cracking and corrosion before the defect diminishes the residual strength of the structure below the acceptable limits. This has replaced the safe life approach and the implementation was executed in the 1970s in order to control the escalation of the maintenance cost due to the safe life approaches, as stated by Koh et. al. (2012). Non-destructive testing was implemented hand in hand with this philosophy as a detection method implemented at certain maintenance intervals or based on the report from the air or maintenance crew on the structural damage (Smith & Wilson, 1985: and Chaumette, 1985). The advantages of damage tolerance are that it can provide life extension through repair and modification and it can optimize structural performance while maintaining safety. The drawbacks are that the implementation on composite structures has opened up a few challenges in applying fracture mechanics for crack propagation due to the fiber-matrix relationship and non-isotropic behavior which can lead to complex and complicated failure modes (Chiu et. al., 1994, and Falzon, 2009).

The improvement of traditional based damage tolerance is being achieved by the use of condition based monitoring. As the damage tolerant approach saves billions of dollars to ensure the safety of the passengers and aircrafts, a new paradigm is required so as to be able to detect damage at its initiation stage. Therefore predictive maintenance can be performed more effectively as is being implemented on helicopter main rotor systems. Both conditioning monitoring and damage tolerant philosophies are consolidated and known as Health Usage and Monitoring Systems (HUMS) (Basehore and Dickson, 1998).

A dependable SHM system is required in order to enhance safety, cost measurements and predictive maintenance. Low (2001) and Hayes (2001) stated that aircrafts are designed with a life between 20 to 25 years. Within this, the maintenance cost is about 12% of the cost of operating the aircrafts. In total, billions of dollars can be saved if improvements in SHM systems are implemented. Achenbach (2009), Farrar and Worden (2007) and Balageas (2002) have concluded that from the design philosophy stated above, the motivations in establishing SHM systems are to:

1. Detect damage or defects at their earliest stage possible. A corrective action can be implemented to contain the damage and repair. This can lead to cost savings and reduced economic impact as a small defect is easy and cheap to repair rather than when it becomes big and uncontrolled.

2. Predict the structure or component behavior with historical data, especially when an extension program is needed to continue operating it. This valuable data can

forecast the limitation of the structure and thus suggested action can be taken to prolong its life.

3. Minimize uncertainties due to un-predictive and predictive events through condition monitoring. Historical data monitoring can provide the behavior of the structures or component, therefore the development of discrepancies can be isolated and remedial action can be taken accordingly.

4. Optimize human capital reliance on technical expertise especially in NDI.

5. Improve the economic consideration by having a more effective assets management approach.

6. Provide an advanced material sensor variation at post repair situation.

13. SHM Roadmap

The aviation sector is a highly regulated body which is governed by the International Civil Aviation Organization (ICAO). Under ICAO, the main regulatory body is mainly divided into the Federal Aviation Administration (FAA) and the European Aviation Safety Agency (EASA) which originated in the United States of America (USA) and European countries respectively. This is mainly due to the aviation and aerospace sectors being dominated by these areas. Malaysia, since it is a British colonial state, adopted the Civil Aviation Regulation (CAR). Although FAA and EASA are separately functionalized, both regulatory bodies share a similar code of aviation regulation, for example Part 25 lists the acceptable design criteria for the damage tolerant design of an aircraft, Part 66 explains matters pertaining to maintenance personnel, Part 145 describes the maintenance activities for the aviation field etc. (Federal Aviation Administration, 2012). In terms of aircraft repair, the regulation stipulated under Part 145, title 14 from FAR, states that any maintenance must be performed using methods prescribed in an approved data sheet such as the Maintenance Manual, the Structural Repair Manual, or the Advisory Circular (AC) 43.13-1B. Inside AC 43.13-1B, it specifies the acceptable methods describing how NDI is to be carried out along with detailed diagrams, checklists and reporting forms.

Commercial aircrafts are designed to be operational for 20-25 years of service and up to 90,000 hours, with the cost to keep the aircraft in operation being around 12% of the total operational cost (Baldwin, 2013). In order to ensure the aircraft's airworthiness, the maintenance of the aircraft has to adhere to a stringent scheduled maintenance plan. The maintenance program was developed through a program called the Maintenance Steering Group (MSG). This group consists of the aircraft manufacturer, the aircraft operator and regulatory personnel who are responsible for creating the maintenance program for any new aircraft and its derivatives (Kinnison, 2004). The maintenance plan is based on a three design methodology which are the safe-life, damage tolerant and condition-based maintenance. Safe-life philosophies are based on hard time components in which replacement is mandatory when it falls due. Economically, this increases the operational cost but safety is kept paramount. The Damage tolerant approach predicts the critical flaw size for a component and a set base of inspection intervals based on prediction to detect and repair the part before failure emerges. Although the philosophies have been implemented for many years, frequent intervals of inspection may ground the aircraft which affects the air operator's revenues. A good maintenance schedule will optimize the inspection intervals, but it requires a handful of NDI inspectors to perform the tasks. Condition based monitoring, however, will monitor the performance of the component in which any initial trigger due to damage will be followed by remedial action. In this case, the cost of inspection, replacement and being grounded are kept low as determined by Dupuy et. al. (2011). Therefore, by implementing the conditioning monitoring principles on an aircraft structure, a more effective damage detection system can be adopted due to new material applications such as advanced composite or hybrid materials whose properties are very different from metallic structures. The system is known as the aircraft structural health monitoring system.

Mahoon (1998) pointed out the NDI acts like a safeguard to prevent catastrophic failure of the aircrafts. For commercial aircraft, the NDI are detailed inside the Maintenance Manual (MM) and the Structural Repair Manual (SRM) created by the aircraft manufacturers to list each of the components to be inspected, the interval of inspection, the type of damage and the recommended methods to be used for the inspections. The NDI inspections are carried out based on flight cycle and hours or upon defects reported by the crews, as shown by Polimeno and Meo (2009) and Rose (2002). Although new techniques are emerging, the inspection has to be carried out on the ground and must be done by certified personnel. Therefore by applying SHM together with NDI, a systematic, yet effective damage detection process can be optimized at the earliest possible time.

A further motivation for applying SHM is to allow the current time based maintenance philosophy to evolve into a more cost effective condition based monitoring approach as is being practiced in rotating machinery, as pursued by many researchers (Wallace et. al., 2004; Ko and Ni, 2005; and Lopez and Sarigul-Klijn, 2010). As has been discussed, the application of SHM will monitor the system response and rectification can be carried out before the whole system fails. In this case, the damage is isolated and total shut down can be minimized. During shut down, the unaffected structure can be operated or can function with minimal effects on the larger system while the damage area is being inspected or repaired. The motivation for applying SHM will lead to the structures having a longer usable time depending on the predictive model combined with regular maintenance checks. This will drastically reduce or eliminate regular inspection by having a longer time between inspections whereby the SHM system can detect damage before a catastrophic failure in time to save the aircraft (Polimeno and Meo, 2009). Besides this, the aircraft operator can gain more opportunity cost by operating the aircraft within the inspection time frame, which may sometimes take more than one day; this can turn out to be a huge investment cost saving in return.

As being investigated by fellow researchers, the use of advanced materials such as fiber composites and hybrid composites poses challenges for the maintenance crew in detecting the damage. (Staszewski et. al., 2009 and Katnam, Da Silva and Young, 2013) Although, the necessary steps have been taken to detect, remove and repair the damage, the uncertainty of the repaired structure is still questionable. In metallic structures, the data for applying damage tolerant and fractured mechanics in order to determine the new usable load are available in standard forms. However, for composite structure damage tolerance and fractured mechanic analysis, the predictive model cannot be directly applied due to constituents, orientation, matrix and hybridization. Current SHM research and findings mainly concentrate on the parent structure being subjected to damage for damage detection purposes (Kessler and Spearing, 2002; Takeda et. al., 2007; Ostachowicz, et. al., 2009; and Konstantin, 2012). The removal and replacement of fiber reinforcement material has to be carried out if any damage extends to the structural layers. Therefore if the structures are equipped with a sensor, there is a need to replace the sensors, and the new reading will either be set to a new baseline or if nothing is changed, the sensor will register the repair area as damage.

Therefore, the motivations in applying the SHM are mainly concentrated on:

1. Optimizing human capital reliance on technical expertise especially in NDI.

2. Economic considerations by having more effective assets management.

3. Advanced material sensor variation in the post repair situation.

There are several inherent difficulties in detecting damage in composite materials as opposed to traditional engineering materials such as metals or plastics. One reason for this is due to their inhomogeneity and anisotropy; most metals and plastics are formed by one type of uniformly isotropic material with very well-known properties. Laminated composite materials, on the other hand, can have a widely varying set of material properties based on the chosen fibers, matrix and manufacturing process as shown by Diamanti and Soutis (2010). This makes the modelling of composites complex, and often non-linear. Another obstacle for many detection techniques is the fact that composites are often a mix between materials with widely differing properties, such as a very good conducting fiber in an insulating matrix. A remaining difficulty is that damage in composite materials often occurs below the surface, which further prevents the implementation of several detection methods. The importance of damage detection for composite structures is often accentuated over that of metallic or plastic structures because of their load bearing requirements. Typically, reinforced plastics are not used in load critical members; since their properties are predictable and they are usually simple and inexpensive to manufacture, they are often designed to be replaceable safe-life parts. Similarly, metals are generally well understood and easy to model, thus they are frequently designed using damage tolerant methodologies. The behavior of composite material, on the other hand, is much less well understood, and an unexpected failure of the composite part could prove catastrophic to a vehicle. New techniques are emerging in order to augment the current techniques to ensure that the probability of detection (POD) within the structure, such as delamination, disbonding and voids, is more effective for modern composite structures, for instance wind generation turbine blades, aircraft pressurized fuselages and civil structures (Amenabar et. al., 2011). Therefore, the development of reliable damage detection methods is critical to maintain the integrity of these vehicles, such that a higher POD can be attained for necessary action to be taken, as investigated by Kopsaftopoulos and Fassois, 2013.

References

[1]   Arby, J.C., Choi, Y.K., Chateauminois, A., Dalloz, B. Giraud G. and Salvia, M. 2001. In-situ monitoring of damage in CFRP laminates by means of AC and Dc measurement, Composite Science and Technology, 61: 855-864.

https://doi.org/10.1016/S0266-3538(00)00181-0

[2]   Benedettini, O., Baines, T.S., Lightfoot, H.W. and Greenough, R.N. 2009. State of the art in integrated vehicle health management, Journal of Aerospace Engineering, 233: 157-170.

https://doi.org/10.1243/09544100jaero446

[3]   Chia, C.C., Lee, J.R., and Park, C.Y. 2012. Radome health management based on synthesized impact detection, laser ultrasonic spectral imaging, and wavelet-transformed ultrasonic propagation imaging methods, Composites Part B: Engineering, 43(8): 2898-2906.

https://doi.org/10.1016/j.compositesb.2012.07.033

[4]   Chia. C.C., Lee, J.R., Park, J.S., Yun, C.Y. and Kim, J.H. 2008. New design and algorithm for an ultrasonic propagation imaging system. Proc Defektoskopie, 4:63–70.

[5]   Chiu, K, Koh, Y.L, Galea, S.C. and Rajic, N. 2000. Smart structure application in bonded repairs, Composite Structures, 50: 433-444.

https://doi.org/10.1016/S0263-8223(00)00110-0

[6]   Garg, D.P., Zikry, M.A., Anderson, G.L. and Stepp, D. 2002. Health Monitoring and reliability of adaptive heterogeneous structures, Structural health monitoring,1(1): 23- 39.

https://doi.org/10.1177/147592170200100103

[7]   Garret, R.C., Peters, K.J. and Zikry, M.A. 2009. In-situ impact induced damage assessment of woven composite laminates through a fiber Bragg grating senor network, The Aeronautical Journal, 113(1144): 357-369.

https://doi.org/10.1017/S0001924000003031

[8]   Giurgiutiu, V. 2008., Structural Health Monitoring with piezoelectric Wafer active sensors, USA: John Wiley.

[9]   Giurgiutiu, V., Zagrai, A. and Bao, J.J. 2002. Piezoelectric wafer embedded active sensors for aging aircraft structural health monitoring, Structural Health Monitoring, 1(41): 41-61.

https://doi.org/10.1177/147592170200100104

[10]   Henderson, I.R., 2002. Piezo Ceramics: Principles and Applications, APC USA: International Inc.

[11]   Herrera, J. M. and Vasigh, B. 2009. A basic analysis of aging aircraft, region of thye world and accidents, Journal of Business and economics Research, 7(5): 121-132.

[12]   Hill, K.O., Fujii, F., Johnson, D.C. and Kawasaki, B. 1978. Photosensitivity on optical fiber waveguides: Application to reflection filters fabrication, Applied Physics Letters, 32: 647-649.

https://doi.org/10.1063/1.89881

[13]   Katsikeros, C.E. and Labeas, G.N. 2009. Development and validation of a strain-based Structural Health Monitoring system, Mechanical Systems and Signal Processing, 23(2): 372-383.

https://doi.org/10.1016/j.ymssp.2008.03.006

[14]   Kesser, S.S. 2002. PhD Thesis: Piezoelectric-based insitu damage detection of composite materials for structural health monitoring systems, in Department of Aeronautics and Astronautics Massachusetts Institute of Technology, Massachusetts Institute of Technology, Massachusetts

[15]   Kousourakis, A., Bannister, M.K. and Mouritz, A.P. 2008. Tensile and compressive properties of polymer laminates containing internal sensor cavities, Composites: Part A, 39: 1394 – 1403.

https://doi.org/10.1016/j.compositesa.2008.05.003

[16]   Lee, J., Chia, C.C., Shin, H.J., Park, C. and Yoon, D.J. 2011. Laser ultrasonic propagation imaging method in the frequency domain base on wavelet transformation, Optics an lasers in Engineering, 49: 167-175.

https://doi.org/10.1016/j.optlaseng.2010.07.008

[17]   Lee, J.R. and Yoon, C.Y. 2009. Development of an optical system for simultaneous ultrasonic wave propagation imaging at multi-points, Experimental Mechanics, 50(7): 1041-1049

https://doi.org/10.1007/s11340-009-9293-y

[18]   Lopez, I. and Klijn, N. S. 2010. A review of uncertainty in flight vehicle structural damage monitoring, diagnosis and control: Challenges and opportunities, Progress in Aerospace Sciences, 46: 247-273.

https://doi.org/10.1016/j.paerosci.2010.03.003

[19]   Michie, C. 2000. Optical fiber sensors for advanced composite materials, Comprehensive Composite Materials, USA: Elsevier

[20]   Mrad, N. 2002. Optical fiber sensor technology: Introduction and evaluation and application: Encyclopedia of Smart Materials, Vol. 2., USA: John Wiley and Sons, 715-737.

[21]   Qiu, L. and Yuan, S. 2009. On development of a multi-channel PZT array scanning system and its evaluating application on UAV wing box, Sensors and Actuator, 15: 220-230.

https://doi.org/10.1016/j.sna.2009.02.032

[22]   Roach, D. 2009. Real time crack detection using mountable comparative vacuum monitoring sensors, Smart Structures and Systems, 5(4): 317-328.

https://doi.org/10.12989/sss.2009.5.4.317

[23] Ryu, C., Lee, J., Kim, C., and Hong, C. 2008. Buckling behavior monitoring of a composite wing box using multiplexed and multi-channel built in fiber Bragg grating strain sensors, NDT & E, 41: 534-543.

https://doi.org/10.1016/j.ndteint.2008.05.001

[24] Salas, K.I. and Cesnik, C.E.S. 2009. CLoVER: An Alternative concept for damage interrogation in structural health monitoring systems, The Aeronautical Journal, 113(1144): 339- 357.

https://doi.org/10.1017/S000192400000302X

[25] Scruby, C.B. and Drain, L.E. 1990. Laserultrasonics—techniques and applications. England: IOP Publishing

[26] Sonatest Inc. 2013. Sitescan D+ Series, Sonatest Limited, Part No. 147359, Issue 2, Product Brochure retrieved on 28 August 2013 from http://www.sonatest.com/products/range/ transducers/probes/single/.

[27] Staszewski, W.J, Mahzan, S. and Traynor, R. 2009. Health monitoring of aerospace composite structures-active and passive approach, Composite Science and Technology, 69: 1678-1685.

https://doi.org/10.1016/j.compscitech.2008.09.034

[28] Valdes S.H.D. and Soutis C. 1999. Delamination detection in composite laminates from variation of their modal characteristics", Journal of Sound and Vibration, 1: 1-9.

https://doi.org/10.1006/jsvi.1999.2403

[29] Verijenko, B. and Verijenko, V. 2005. A new structural health monitoring for composite laminates, Composite Structures, 21: 315-319.

https://doi.org/10.1016/j.compstruct.2005.09.024

[31] Wang, S., Kovalik, D.P. and Ching, D.D.L. 2004 Self sensing attained in carbon fiber polymer matrix structural composites by using the interlaminar interface as a sensor, Smart Material Structure, 13: 570-592.

https://doi.org/10.1088/0964-1726/13/3/017

[32] White, C., Herszberg, I. and Mouritz, A.P. 2009. Structural Consequences of Sensor Cavities In Scarf Repairs, Materials Forum. 33: 427-434

[33]   Whittingham, B., Li, H.C.H., Herszberg, I. and Chiu, W.K. 2006. A disbond detection in adhesively bonded composite structures using vibration signatures Composite Structures, 75: 351–363.

https://doi.org/10.1016/j.compstruct.2006.04.055

[34]   Zhang, H., Schulz M.J. and Feruson, F. 2002. Structural health monitoring using transmittance functions, Mechanical Systems and Signal processing, 2: 357-378.

Chapter 3

A preliminary study of SHM compared with NDT

[1]K. D. Mohd Aris, [2]A.Hamdan, [2]F. Mustapha

[1]Unversiti Kuala Lumpur Malaysian Institute of Aviation Technology, 2891, Jalan Jenderam Hulu,43800 Dengkil, Selangor Malaysia

[2]Department of Aerospace Engineering, Universiti Putra Malaysia, 43400 Serdang, Selangor, Malaysia

Keywords

Structural Health Monitoring (SHM), Non-Destructive Inspection, Condition Monitoring, Composite Material, Piezoelectric

Abstract

Structural health monitoring (SHM) systems are still new in the aircraft system technology. There are many opportunities to employ the technology on composite structure especially in the aircraft industry. The application of composite in aircrafts can increase the demand and awareness to use the SHM concept for structural integrity purposes. The initial testing shows the advantages of using the SHM system in detecting embedded damages of a Carbon Fibre Reinforced Polymer (CFRP) panel when compared to existing non-destructive inspection (NDI) methods. The NDI technique itself is labor intensive and very costly. Therefore, the use of SHM is important for augmenting the NDI techniques. Since PZT smart sensors are widely available, cheap and lightweight, they may become a system of choice when implementing a SHM system to monitor the structure's status in undamaged and repaired conditions.

Contents

1. Structural health monitoring sensor

Current approaches to ensure aircraft structural integrity for metallic components rely on a safe life and condition monitoring philosophy. However, when these approaches are combined with a continuous damage detection system, the whole can be called a health and usage monitoring system (HUMS) for rotary aircraft or a structural health monitoring systems (SHM) for aircraft structural components (Qiu and Yuan, 2009; and Salas and Cesnik, 2009). The integration of the monitoring systems for a system and structure are known as Integrated Vehicle Health Management (IVHM) which implements an advanced prognostic and health management strategy which enables continuous monitoring and real time assessment of the vehicle functional health, predicts remaining useful life or faulty or near failure components, and uses this information to improve operational decisions (Benedettini, et. al., 2009). This approach can lead to a reduction of NDI on components, and in terms of a composite structure any propagation of insidious damage can be detected at its initiation stage. The introduction of SHM-based component systems on composite structures may improve in certifying these structures to be used on aircraft structures. This is due to the different configuration and material hybridizations of the constituents from different manufacturers. Using a smart sensor concept, the damage and its propagation in the airframe and other structural life related components can be continuously monitored on board the aircraft to provide real time damage assessment. Thus, it can prolong the maintenance intervals of some components and reduce the through life cost. The SHM basic components consist of a sensor, data/information processing, diagnostic and prognostic algorithm development and data dissemination and data storage.

There are two types of sensory systems which are passive and active as specified by Staszewski et. al. (2009). The passive sensory smart structure contains only sensors and electronics which are capable of processing the sensor data with structural condition information. The active system contains both sensors and actuators in which there is a similar capability to the passive system, but with the on demand capability to interrogate the structure at arbitrary times.

2.   PZT sensors

The characteristics of the piezoelectric effect on a material were firstly discovered by Jacques and Pierre Currie in 1880 (Henderson, 2002; and Giurgiuttu, 2008). Piezoelectric materials, such as lead zirconate titanate [Pb(ZrTi)O$_3$] or PZT, become electrically polarized when they are subjected to a mechanical force, and mechanically in tensile or compression when an electrical input is induced. The displacement is very small but the application can be found in cases from small electrical appliances to advanced systems such as in aerospace. The properties of PZT are very stable in the range of -22°C to +155°C. It can exhibit most of the characteristics of ceramics such as high elastic modulus, brittleness and low temperature strength. It is manufactured in thin plates, strips or fibers suitable either for surface boning or embedding. The material is isotropically behaved and assumed transversely isotropic in the normal to the poling direction. The advantages of using PZT sensors are their lightness, good power consumption and sensitivity to small strain and acceleration. Moreover, the main benefit in using PZT sensors is the wide utilization of detection techniques by simply altering the actuating signals either in time-based or frequency based analyses. Significant findings by Zhang et. al. (2002) and Valdes and Soutis, (1999) showed that the optimal frequency range to actuate the structure was between 10-20kHZ. However, to collect the data, a frequency of less than 5kHz was practical. A frequency sweep between 8-14kHZ was used to induce vibration on the structure with respect to the increase of delamination size.

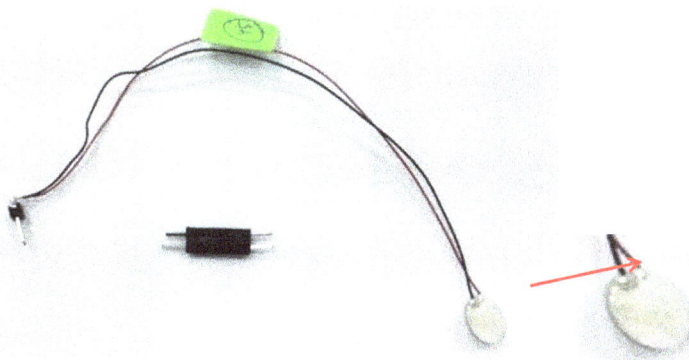

Figure 1 PZT sensor

3. Fiber bragg grating (FBG)

FBG is a type of distributed Bragg reflector constructed in a short segment of optical fiber that reflects particular wavelengths of light and transmits others. It is achieved by creating a periodic variation in the refractive index of the fiber core which generates a specific dielectric mirror of a particular wavelength (Ryu et. al., 2008). The Bragg created by laser illumination acts as an optical filter for the desired wavelength or wavelength specific reflector. It was first discovered and demonstrate by Hill in 1978 and perfected by Meltz in 1989 using a UV side writing technique through laser inscription (Hill et. al., 1978). There are two major classifications of optical fibers, which are single and multi-mode fibers. A single mode will have a core diameter of 10 μm whereas the latter has a diameter between 50 μm to 100 μm. The fibers are capable of withstanding a strain up to 5% but any micro crack in the cladding will lead to degradation of the fiber strain to failure. Moisture will also degrade the performance whereby the buffer materials enclosed and protected the inner layer of the optical fiber. An embedded optical fiber will have an operational temperature range between 120-180°C, especially for the CFRP component in which the curing temperature is around a similar temperature range. The detection parameters by optical fibers consist of pressure, strain, temperature, and chemical as studied by Mrad (2002) and Michie (2000). The advantages of using optic sensors are good spatial resolution, immunity to electromagnetic interference, good fatigue and durability, provision of distributed or point sensing, and operation at a wider temperature range. The disadvantages of using this system are that, at present, the cost is too high for the complete system and the bulky instrumentation which requires a space and weight penalty.

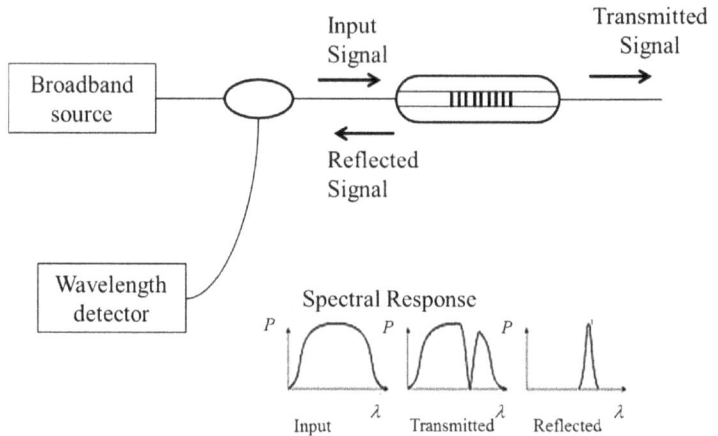

Figure 2 Typical schematic diagram for FBG System (Fu et. al., 2009).

4. Comparative vacuum monitoring (CVM)

CVM technology is an embedded system which measures the differential pressure between fine galleries containing a low vacuum alternating with galleries at atmospheric pressure in a manifold (Kousourakis et. al., 2008). For an embedded SHM system, vacuum monitoring is applied to small galleries that are placed adjacent to a second set of galleries maintained at atmospheric pressure. If there is no flaw, the low vacuum remains stable at the base value. If a flaw exists, the air will flow from the atmospheric galleries through the flaw into the vacuum galleries. A crack in the material beneath the sensor will allow leakage resulting in detection due to the rise of the monitored pressure. For an external surface, it can monitor the crack initiation, propagation and corrosion by the same principle of pressure rise within the galleries, detection of cracking, fatigue and corrosion detection and initiation. One of the advantages of the system is that it does not require any electrical excitation or contact of the sensor, as stipulated by Roach (2009). The gallery can increase fatigue resistance as shown by White, Herszberg and Mouritz (2009). However, the main consideration concerns size; due to the fact that the introduction of the systems inside the aircraft structure can lead to a certification issue, in which the size of the sensor may lead to residual delamination.

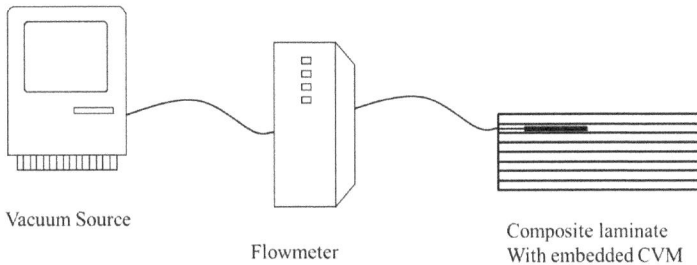

Vacuum Source

Flowmeter

Composite laminate
With embedded CVM

Figure 3 Typical Schematic diagram of Comparative Vacuum Monitoring system (Kousourakis, Mouritz and Bannister, 2006)

There are two factors which need to be considered in order to obtain useful SHM measures. The first is the system level consideration which encompasses the features to be extracted from the data for SHM assessment, active and/or passive sensing, operational & environmental factors and the constraints of the sensor system. Secondly there is the data acquisition component consideration, which is involved with the sensitivity and bandwidth of the sensors used, resolution of detectability, cross axis sensitivity in which is the arrangement effectiveness of the sensors, resonance sensitivity and sensitivity to unwanted readings.

5. Optical laser ultrasonic propagation

A new, emerging technique is being investigated by Lee, et. al. (2011) known as the Ultrasonic Propagation Inspection (UPI). It is a noncontact SHM system which operates by scanning the surface irregularities using a laser. Unlike the contact SHM methods, which exploit the on surface response, this system is placed at a standoff distance and consists of a diode-pumped solid state laser (DSSL), a galvanomotorized Laser Mirror Scanner (LMS), an ultrasonic sensor, a signal conditioning subsystem, a digitizer, and a computer for hardware control and results processing, as shown in Fig. 4. A full description of the system setting can be reviewed from the work of Lee and Yoon (2009). The DSSL is a Q-switched Nd:YAG diode-pumped solid-state laser with a pulse repetition. The LMS was used to deflect laser pulses toward a target structure with two galvano-motors. Each of the galvano-motors has a laser mirror that is capable of rotating the mirrors orthogonal to each other. Therefore, the LMS can manoeuver the laser impingement point rapidly within a two-dimensional (2-D) scan field. During the scan, the LMS reflects the laser pulses such that the pulses can go vertically downward (V-axis), step horizontally (H-axis), and then scan vertically upward. The output of the DSSL is a series of laser pulses which generates a grid of laser impingement points on the target structure. The pitch of the scanning grid, Δ, is constant in both directions of the V- and H-axes, which is determined by adjusting the standoff distance of the LMS from the target structure, the PRF of the DSSL, and the angular speed of the galvano-motors. The pulses emitted by a DSSL with a pulse repetition rate in the kilohertz regime could be controlled due to the sufficient speed of the galvano-motors. When the laser beam impinges on the surface of the target, an ultrasonic wave is created at the affected point. The mechanism of the generation of an ultrasonic wave in solid media was well documented in Scruby and Drain,(1990). The ultrasonic wave propagates through the target structure and reaches the ultrasonic sensing point. Ultrasonic reception can be realized using a contact or non-contact ultrasonic sensor. Sensor installation for the UPI system is simple because sensing at a fixed point is sufficient for the UPI system. The sensing point of the ultrasonic wave can be on the same side, or on the opposite side relative to the laser scan. This feature is important because access to both surfaces of the target structure is not always possible in IPQC, NDE, and SHM. The signal from the sensor is amplified and filtered by a signal conditioning subsystem. Filtering is not needed for the frequency selection, but is needed for the reduction of random ambient noise. The signal is then sampled for N number of data at a sampling time interval T using a digitizer plugged into the computer (Chia et. al. 2008).

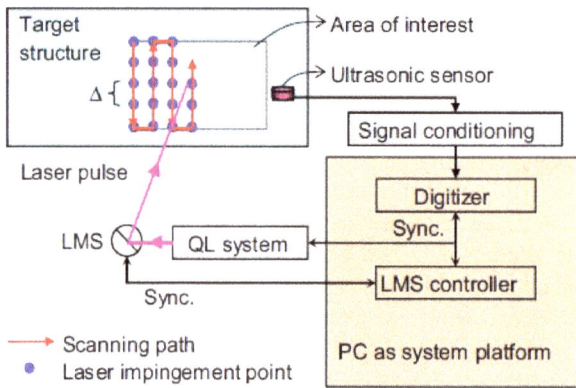

Figure 4 Schematic diagram of the UPI system (Chia, Lee and Park, 2012)

Although the system does not impose any significant changes on the structural integrity compared to other SHM systems, the equipment is too bulky and requires higher computing power to compensate for the processing speed of the computing equipment due to the laser pulse.

6.   Other sensor systems

Since there are no live SHM sensors being used aboard commercial aircraft, systems that utilize a combination of sensors and systems are also under investigation. *Stanford Multi Actuator Receiver Transduction (SMART)* is a combination of composite structures for load bearing and sensors and actuators for information acquisition, processing and control whereby the layer is embedded within the structure (Garg et. al., 2002). The advantage of the system is that it covers more area compared to a single PZT set due to its integrated circuitry.   However, by embedding the sensors, the structural integrity need to be strengthened due to the strength of the structure is related to the size of the affected sensor removed and replaced on the damage structure. *Metastable Ferrous Alloy Inserts (MFAI)* is a strain memory alloy which is austenitic in nature and is embedded within the laminate. Although the strain memory is austenitic at room temperature, once strained it becomes permanently martensitic. The change of the structure will have an effect on the strain which can be detected by the use of a non-contact sensor (Verijenko and Verijenko, 2005). The advantage of the system is that it uses strain energy from the sensor to differentiate between a health state and a damage state. However, the fabrication process must be done carefully so that the sensor does not cause damage during the fabrication and embedment of the senor within the laminates. Carbon –fiber reinforced polymer uses the material itself as the conductive path to measure the global electrical resistance of a structure (Arby et. al. 2001; and Wang et. al., 2004). Thus, any damage occurring internally will have a direct effect on the resistance. The method proves to be successful in determining the health state of the structure, but the localization of the damage is hardly justified based on the resistance only.

 Sensor size and detection play an important role where the application confined to aerospace structure is concerned. The higher detectability with the smallest size of sensor is the most desirable since it can save the structural weight. However, each sensor has its limit where the above conditions are concerned as shown in Table 1.

Table 1 Sensor Capability. (Kesser, 2002)

Type of sensor	Application	Size of sensor (mm)	Size of Damage(mm)	Sensor Power(Watt)
PZT sensor	Entire Plate	3-100	6-10	0.01-15
Optical Fiber	Sensor Area	0.2-3	10-30	0.01-0.1
Eddy Current	Sensor Area	11-100	0.06-1	0.01-1
Acoustic Emission	Half Plate	3-100	1-8	0.01-0.02
Modal Analysis	Entire Plate	0.3-100	10-100	0.01-10

The capability of damage detection of the sensors also varies due to the above conditions. Ideally, a single sensor should be able to detect various types of damage with a single application. Unfortunately, due to this limitation a compressive type of sensor needs to be used for its specific usage. Furthermore, the regulatory powers do not want the detection of damage to be obligated to a single type of sensor. At least one other different type of sensor must work hand in hand to detect the damage in order to increase the effectiveness of the damage detection process. Table 2 shows various capabilities drawn from various ongoing works on the SHM system.

Current aircraft are mainly made from aluminum alloy in the main structural parts. However, this is going to change as the technology of advanced composite materials has superior improvements over the metallic structures. Some of the sensors are being introduced to monitor the structural integrity of aging aircraft and the current operational aircraft in order to increase the safety, reliability and aviation economics as shown in the research of Herrera and Vasigh (2009), Chiu et. al. (2000), Giurgiutiu et. al. (2002) and Lopez and Sarigul-Klijn (2010). The ongoing research will benefit the future Boeing 737 and Airbus A350 replacement program, but the date of introduction is still unknown. These aircraft will undergo a technology leap in terms of structure, system, powerplant, avionics, in-flight entertainment systems and cabin atmosphere systems.

Table 2 Sensor Utilization (Kesser S.S., 2002; Garret et. al. 2009; Verijenko and Verijenko, 2005; Katsikeros and Labeas, 2009; and Kousourakis et. al.,2008)

Sensor types		Damage detection method	Area Of detection	Structure materials	Mode of detection
Piezoelectric sensors	PZT	Delamination Impact	Local & Global	Composites & Metallic	On-line and/or offline
Fiber Bragg Gratings	FBG	Loads Impacts Delamination	Local	Composites & Metallic	On-line
Foil Strain Gauges	SG	Cracks	Local	Metallic	Off line
Comparative Vacuum Monitoring	CVM	Cracks Corrosion Debondings	Local	Composites & Metallic	Off line

7.   Preliminary comparison between NDI and PZT

The preliminary experimental analysis for the functional assessment of the relevant specimens studied is presented in this chapter. The chapter is subdivided into the fabrication of the specimen and the comparison between a typical NDI and PZT sensor's response.

8.   Specimen Fabrication

Preliminary study on SHM conducted using Carbon Fiber Reinforced Plastic (CFRP) materials. Specimen was fabricated for the intended testing which were specifically for the panel testing. A prefabricated CFRP panel was used for the aircraft structure component. Uni-directional (UD) pre-impregnated (pre-preg) Hexply® M10/38%/UD300/CHS/460mm CFRPs from Hexcel Corps were used for the various types and sizes of the panels. As for the pre-impregnated CFRP materials, they were cured using heat assisted temperature equipment known as a hot bonder. The CFRP panels was fabricated to $[0/90]_{S2}$ orientation in order to produce a balanced, symmetrical, flat monolithic structure.

Table 3 Distribution of the experiments

No.	Type of Test	Materials	Final Size	Orientation
1	Comparison with various NDI techniques	UD pre-preg (CFRP)	300mm by 300mm	0°

The panels were divided into two sections which are the undamaged and damaged panels. The undamaged specimens went through a one-step curing process. Next, some of the undamaged panel was subjected to a simulated damage and repair. Once the repair was completed, the specimens were examined in individual testing.

Panels 1 originated from Hexply® M10/38%/UD300/CHS/460mm manufactured by the Hexcel Corporation (Hexcel Corporation, 2005). The material was a pre-impregnated carbon fabric of unidirectional (UD) origin. The curing was done at an elevated temperature using a hot bonder machine supplied by Heatcon Inc. from the HCS9000B series.

The initial test was conducted on a simulated, embedded damage defect in CFRP panel. The panel was cured by heat assisted curing using a hot bonder and consolidated using the vacuum bag system. Next, the panel was divided into 5 by 5 grids of 10mm by 10mm size each. Two simulated defects were embedded at the sublaminate layer at predetermined grid locations.

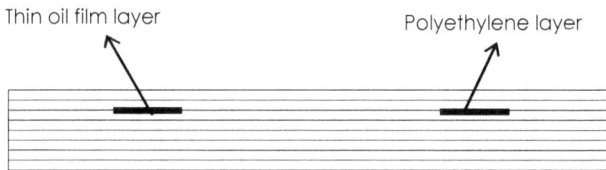

Thin oil film layer   Polyethylene layer

Figure 5 Preparation for the embedded damage panel.

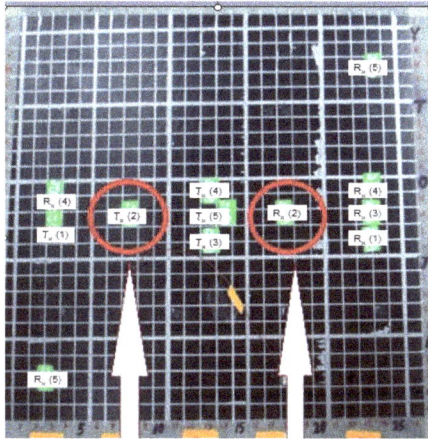

Figure 6 PZT placement and defect location on the CFRP panel

9. Non Destructive Inspection

Only a few NDI methods were used, such as the tap test, the ultrasonic A-scan and radiography for the preliminary study to compare with the smart intelligent SHM sensor. These techniques complied with the current assessment and inspection used in the present aero-structure maintenance and inspection schedules. All testing was conducted in the laboratory except for the radiography inspection, which was carried out in an isolated place due to the danger of radiation exposure during the test.

10. Tap test (Wichitech – RD3 Tap Hammer)

The tap test was conducted using an RD3 digital tap hammer from Wichitech Inc. (RD3 Digital electronic tap hammer, Wichitech Industries, Inc., Product Brochure). This NDI method was developed to detect voids, degradation, and delamination in composite structures. It uses the echo sensed from the hammer which is translated into numerical form on the display unit during the tapping. It supplements the subjective tonal discrimination of the operator with a quantitative, objective numeric readout that can be correlated to delamination in the structure. A reference tapping within the adjacent area was needed in order to detect anomalies such as disbonding or delamination. This is known as local referencing. The tap hammer was used by lightly beating the panel on each grid. The results were shown in a coloured matrix for the anticipated simulated defects.

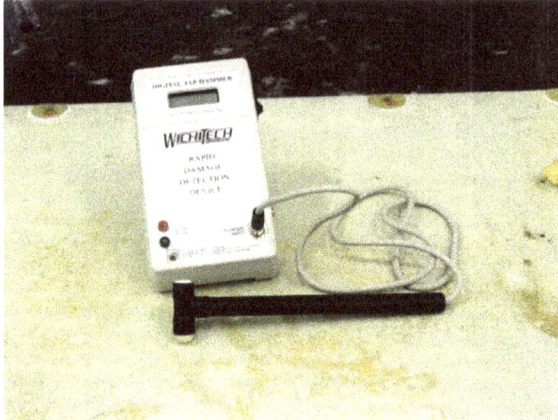

Figure 7 Tap Test from Whichitech RD3

By tapping at the same phase on each gradient or zone, the results were displayed in a colour gradient from red to white, the former being the value near to the reference and the latter indicating the value near the defect. The result showed a reliable acknowledgement of the location of the simulated defects. The variation of the sound produced for the damages were in the area between 550 ~ 599 which were in grids C2 and C4. This result compared with the embedded defects. However in a real situation, the structure can only be inspected while the panel is isolated and in aircraft usage; the aircraft needs to be on the ground and it heavily relies on a manpower task.

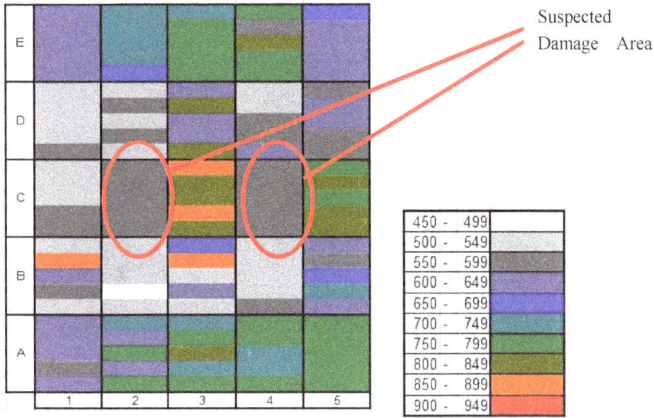

Figure 8 A colour scale produced by the RD-3 Digital Tap test with classification from the tap test feedback.

11. Ultrasonic inspection – A-scan

The ultrasonic inspection was conducted using a Sitescan D-10 ultrasonic Flaw Detector from Sonatest (Sitescan D+ Series, Sonatest Limited, Part No. 147359, Issue 2, Product Brochure). It uses high-frequency sound waves transmitted onto a material to detect imperfections or to locate changes in the material properties using a pulse echo method. The sound was introduced into a test object through knocking at its insulated probe, and reflections (echoes) were produced from internal imperfections or the part's geometrical surfaces. The expanding waves were detected by the receiver probe sensor. A 5 Mhz probe (http://www.sonatest.com/products /range/transducers/probes/single/downloaded on 20/11/2013) was selected to be used for the inspection due to the nature of the anistropic properties of the materials, the thin plate configuration and the penetration properties. A thin layer of couplant is applied on the contact surface of the probe in order to transfer the sound wave from the sensor to the specimen. The inspection was carried out by scanning the panel with the probe across the grid.

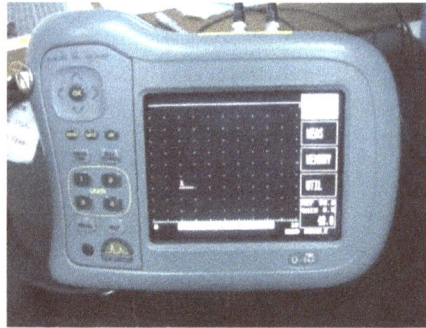

Figure 9 Ultrasonic A-scan from Sonatest Sitescan D-10

Before the actual reading was taken, a reference reading was acquired by using the reference block for calibration purposes and this showed the respective response accordingly. The isotropic properties of the reference metallic structure provided a clear front wallback wall wave generation and the simulated defect position is as shown in the figure below. The presence of the defect will appear in between these ranges.

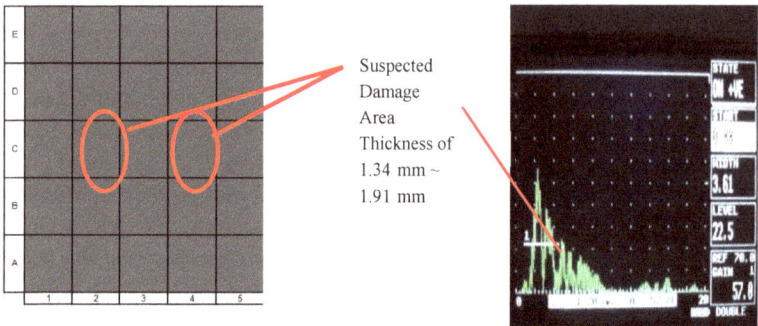

Figure 10 Results from the Sonatest D-10 Ultrasonic Reading on the CFRP Panel.

For the CFRP panel, the thickness was measured and found to be at an average of 2.60mm. By using the same probe and procedures for the CFRP panel NDI, the result for frontwall and backwall are hard to determine, and it required assistance from an approved

NDI holder to determine the result to set the time of travel, sound medium and frequency for the probe. In addition, the un-isotropic properties of the CFRP panels (due to the fiber orientation, fiber-matrix composition and consolidation) may lead to the scattering of the sound from the transducer. From the result, the defect location was indicated by distance between the surface and the defect location, which resulted in being between 1.32 and 1.2 inch from the inspected surface which is the embedded position of the defects. The result showed that the technique was able to detect damage on the CFRP panel although the findings were deviated.

12. Radiography inspection

The test was carried out by exposing the specimen to material isotopes. The source was from the Iridium 192 radioactive isotope (Halmshaw, 1991). The isotopes were placed close to the material to be inspected and they passed through the material and were captured on film. The film was processed and the image was obtained as a series of grey shades between black and white. Gamma sources have the advantage of portability which makes them ideal for use in site working. Due to its hazardous nature and safety consideration, the inspection of the plate was conducted in an enclosed area and by approved personnel. Figure 11 shows the placement of the isotopes with respect to the specimens and the film.

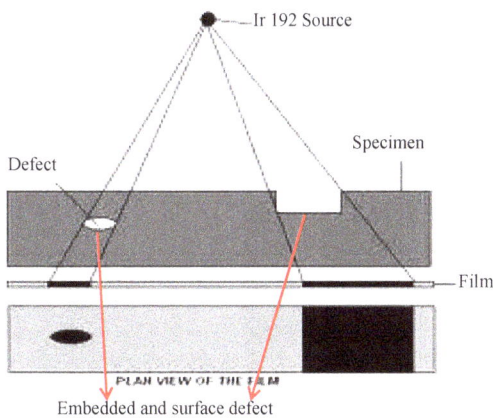

Figure 11. Ir 192 typical procedure

The testing was conducted in an isolated and controlled facility due to the required safety precautions when using the Iridium 192 source. The exposed radioactive source was targeted on the CFRP panel during the inspection. The average thickness of the CFRP panel was 2.60mm. The result showed no indication of the simulated defects on the film generated after the panel was exposed to the radioactive source. There was no contrast on the developed film on the C1 to C5 section. This result suggests that radiographic inspection is not suitable for detecting damage embedded within the plies.

Figure 4.12. Result from Ir-92 Radiography film

13. PZT sensor

The smart APC-850 PZT sensors from APC International were placed to react as the transmitter or receiver from 100mm apart to 500 mm apart. The locations were prescribed, $R_x(n)$ was the receiver, $T_x(n)$ was the transmitter, and n is the reading iteration during the data acquisition. The sensors were connected to both the TG1010A TTi oscilloscope and the GDS-2104 GW Instek wave generator for delivering and receiving the signals as shown in Figure 12. A total of 10 readings were collected for each configuration. Before the testing started, all PZT sensors were inspected for any defects by measuring the dimension, visual inspection on the soldered terminal and continuity check for any unseen disconnection.

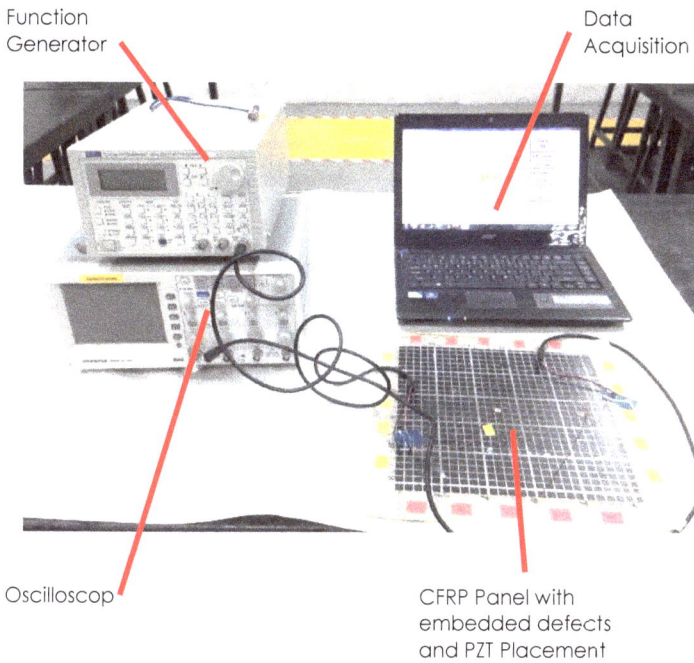

Figure 12 Hardware configuration for data acquisition process.

One set of the PZT sensors was used on this validity inspection to compare with the NDI results. Since a polymer double sided tape was used, a similar PZT can be salvaged for the test as long as proper care is taken to avoid any breakage or damage to the sensors and connectors. One PZT was placed as the actuator and the other as the receiver. The correct response was achieved by selecting the best time, frequency and voltage scales from the oscilloscope to attain minimum noise and interference. The actuating signals were set at the optimum phase as per Table 4.

Table 4 Setting Parameter for oscilloscope and wave generator

Function Generator		Oscilloscope	
Frequency	50kHz	Power	DC
Burst	5	Time division	25ms
Time generation	3ms at 333 Hz	Ch1 Voltage	5V
Voltage Peak to peak	10V	Ch2 Voltage	20mV
Phase angle	0º	Sensor distance	100mm and 500mm

Upon the parameters were set, the actuating signal was examined to ensure proper output was retrieved during the testing. These set up parameters were also used for the rest of the testing in the following chapters. The burst signal is shown in Figure 13.

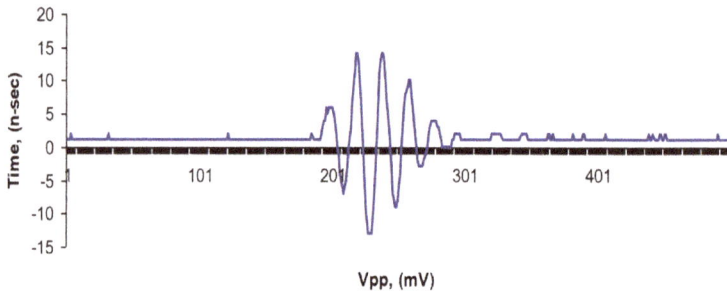

Figure 13 Input signal from APC 850

There were five different PZT positions placed on the grid. At each position, the two PZT sensors acted as an actuator and receiver across the simulated damage. The wave responses are shown side by side. Fifty readings were taken for each configuration. At this stage, the results of interest were the response of the Voltage peak to peak (Vpp) which was assumed to have an effect with respect to defects and distance between the two PZT sensors. The results for the response signals are shown from Figure 14 to Figure 17 for any configuration tested.

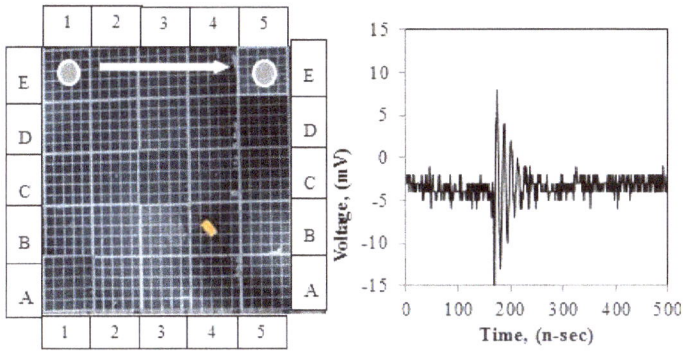

Figure 14. CFRP grid between E1 and E5

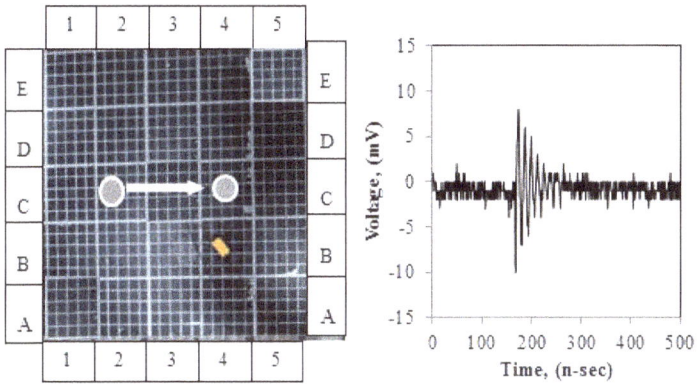

Figure 15 CFRP grid between C2 and C4

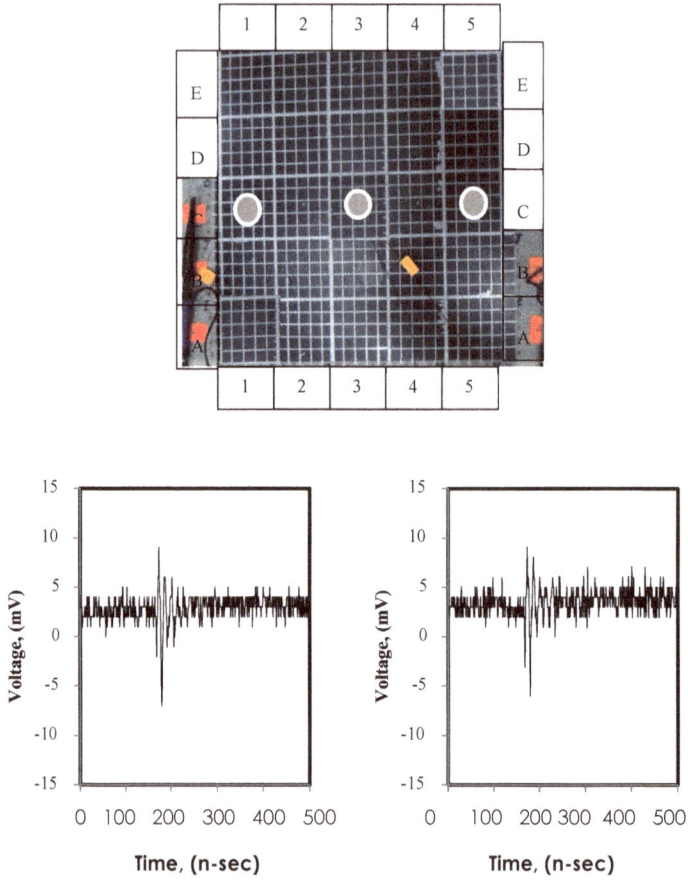

Figure 16. CFRP grid between C1-C3 and C5-C3

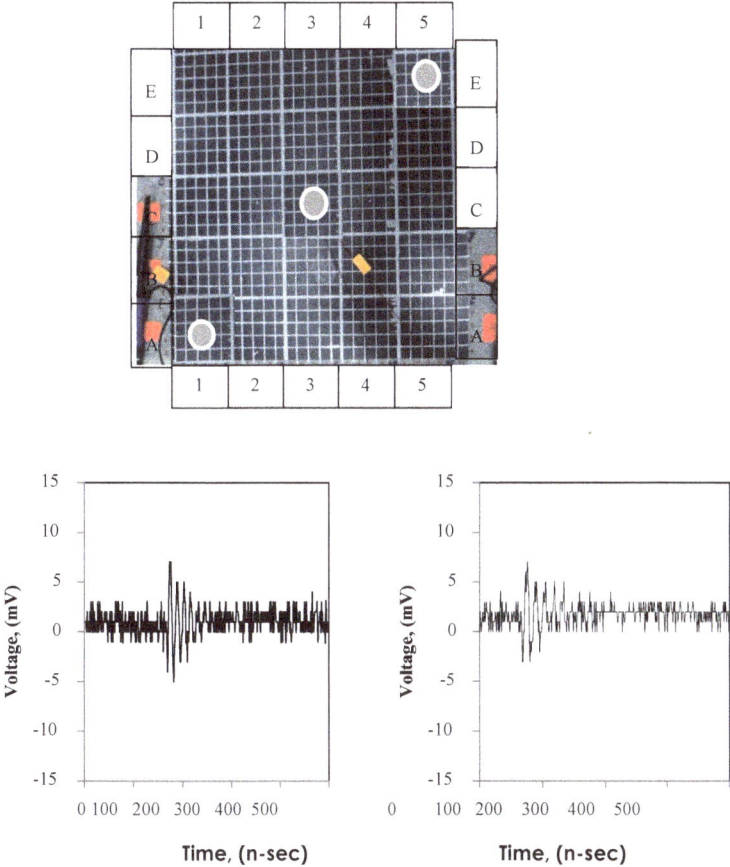

Figure 17 CFRP grid between C3- A1 and C3-E5

The values of the voltage peak to peak (Vpp) retrieved were calculated and compared. Any indication of a different Vpp value showed a hidden anomaly underneath the laminated skins. The Vpp values were calculated by taking the absolute value of the maximum voltage to the minimum voltage. The results are depicted below.

Table 5 PZT Placement Configuration

	Tx Location	Rx Location	Tx~Rx Distance, (mm)	RxVpp, (mV)	Tx~Rx Position
Config 1	E1	E5	200	23	Inline
Config 2	C2	C4	100	18	Inline
Config 3	C3	C1	100	16	Inline
Config 4	C3	C5	100	15	Inline
Config 5	C3	A	150	12	Diagonal
Config 6	C3	E5	150	10	Diagonal

The results from Table 5 show a significant higher reading on Config 1 which is the wave perturbated on the undamaged area. The readings from the other configurations show a reduction of Vpp values once the wave has travelled on a damaged area. This can lead to the surface wave which can detect surface irregularities at the sublaminate layers. At 100mm apart, the PZT sensors at Configs 2, 3 and 4 show a higher reading than at Configs 5 and 6. This shows that the surface wave decimated over distance.

14. Discussion

From the experiment, the tap test machine has the ability to detect the damage area by mapping the findings. The tap test was adequate to detect the existence of damage but it cannot perform the localization of the damage within the laminate. The ultrasonic A scan inspection performed well in detecting the simulation damage. The penetration of the ultrasonic wave was limited to the nearest damage possible from the probe. It can detect the delamination at the sub surface layer by the different thickness from the reference spots. However, it cannot localized and pinpoint the damage area. The use of radiography is very limited on a thin structural plane. The oil traces and FOD cannot be seen on the developed film suggesting that both were bonded to the laminate. The use of a PZT smart sensor transducer can detect the damage by the different signal received. The equipment was turned on and any anomalies could be detected on-line. This suggests that live monitoring can be done to support the continuous structural health monitoring of the aircraft. The development of an SHM system is still being studied. Further investigation will focus on the data gathering, generic algorithm from the actual aircraft component which can lead to an efficient system prognosis.

Summary

Structural health monitoring using PZT sensors to detect anomalies for repair conditions has been studied. The initial testing shows the advantages of using the SHM system in detecting embedded damages of a CFRP panel when compared to existing NDI methods. The use of PZT sensors for SHM systems provided a continuous active monitoring on the CFRP panels. It was also found that the radiography NDI was not suitable for thin materials.

In conclusion, the above recommendations shows there are many opportunities for applying SHM technology on aircraft structures. SHM systems are still new in aircraft system technology. The use of composite materials has expedited its demand and can hasten its maturity to be introduced on new generation aircraft due to the autonomous requirement for detecting damage at the earliest time possible.

The NDI technique itself is labor intensive and very costly. Therefore, the use of SHM is important for augmenting the NDI techniques. Since PZT smart sensors are widely available, cheap and lightweight, they may become a system of choice when implementing the SHM system to monitor the structure's status in the undamaged and repaired conditions.

References

[1]   Arby, J.C., Choi, Y.K., Chateauminois, A., Dalloz, B. Giraud G. and Salvia, M. 2001. In-situ monitoring of damage in CFRP laminates by means of AC and Dc measurement, Composite Science and Technology, 61: 855-864.

https://doi.org/10.1016/S0266-3538(00)00181-0

[2]   Benedettini, O., Baines, T.S., Lightfoot, H.W. and Greenough, R.N. 2009. State of the art in integrated vehicle health management, Journal of Aerospace Engineering, 233: 157-170.

https://doi.org/10.1243/09544100jaero446

[3]   Chia, C.C., Lee, J.R., and Park, C.Y. 2012. Radome health management based on synthesized impact detection, laser ultrasonic spectral imaging, and wavelet-transformed ultrasonic propagation imaging methods, Composites Part B: Engineering, 43(8): 2898-2906.

https://doi.org/10.1016/j.compositesb.2012.07.033

[4]   Chia. C.C., Lee, J.R., Park, J.S., Yun, C.Y. and Kim, J.H. 2008. New design and algorithm for an ultrasonic propagation imaging system. Proc Defektoskopie, 4:63–70.

[5]   Chiu, K, Koh, Y.L, Galea, S.C. and Rajic, N. 2000. Smart structure application in bonded repairs, Composite Structures, 50: 433-444.

https://doi.org/10.1016/S0263-8223(00)00110-0

[6]   Garg, D.P., Zikry, M.A., Anderson, G.L. and Stepp, D. 2002. Health Monitoring and reliability of adaptive heterogeneous structures, Structural health monitoring,1(1): 23- 39.

https://doi.org/10.1177/147592170200100103

[7]   Garret, R.C., Peters, K.J. and Zikry, M.A. 2009. In-situ impact induced damage assessment of woven composite laminates through a fiber Bragg grating senor network, The Aeronautical Journal, 113(1144): 357-369.

https://doi.org/10.1017/S0001924000003031

[8]   Giurgiutiu, V. 2008., Structural Health Monitoring with piezoelectric Wafer active sensors, USA: John Wiley.

[9]   Giurgiutiu, V., Zagrai, A. and Bao, J.J. 2002. Piezoelectric wafer embedded active sensors for aging aircraft structural health monitoring, Structural Health Monitoring, 1(41): 41-61.

https://doi.org/10.1177/147592170200100104

[10]   Henderson, I.R., 2002. Piezo Ceramics: Principles and Applications, APC USA: International Inc.

[11]   Herrera, J. M. and Vasigh, B. 2009. A basic analysis of aging aircraft, region of thye world and accidents, Journal of Business and economics Research, 7(5): 121-132.

[12]   Hill, K.O., Fujii, F., Johnson, D.C. and Kawasaki, B. 1978. Photosensitivity on optical fiber waveguides: Application to reflection filters fabrication, Applied Physics Letters, 32: 647-649.

https://doi.org/10.1063/1.89881

[13] Katsikeros, C.E. and Labeas, G.N. 2009. Development and validation of a strain-
      based Structural Health Monitoring system, Mechanical Systems and Signal
      Processing, 23(2): 372-383.

https://doi.org/10.1016/j.ymssp.2008.03.006

[14] Kesser, S.S. 2002. PhD Thesis: Piezoelectric-based insitu damage detection of
      composite materials for structural health monitoring systems, in Department of
      Aeronautics and Astronautics Massachusetts Institute of Technology,
      Massachusetts Institute of Technology, Massachusetts

[15] Kousourakis, A., Bannister, M.K. and Mouritz, A.P. 2008. Tensile and
      compressive properties of polymer laminates containing internal sensor cavities,
      Composites: Part A, 39: 1394 – 1403.

https://doi.org/10.1016/j.compositesa.2008.05.003

[16] Lee, J., Chia, C.C., Shin, H.J., Park, C. and Yoon, D.J. 2011. Laser ultrasonic
      propagation imaging method in the frequency domain base on wavelet
      transformation, Optics an lasers in Engineering, 49: 167-175.

https://doi.org/10.1016/j.optlaseng.2010.07.008

[17] Lee, J.R. and Yoon, C.Y. 2009. Development of an optical system for
      simultaneous ultrasonic wave propagation imaging at multi-points, Experimental
      Mechanics, 50(7): 1041-1049

https://doi.org/10.1007/s11340-009-9293-y

[18] Lopez, I. and Klijn, N. S. 2010. A review of uncertainty in flight vehicle structural
      damage monitoring, diagnosis and control: Challenges and opportunities, Progress
      in Aerospace Sciences, 46: 247-273.

https://doi.org/10.1016/j.paerosci.2010.03.003

[19] Michie, C. 2000. Optical fiber sensors for advanced composite materials,
      Comprehensive Composite Materials, USA: Elsevier

[20] Mrad, N. 2002. Optical fiber sensor technology: Introduction and evaluation and
      application: Encyclopedia of Smart Materials, Vol. 2., USA: John Wiley and Sons,
      715-737.

[21] Qiu, L. and Yuan, S. 2009. On development of a multi-channel PZT array
      scanning system and its evaluating application on UAV wing box, Sensors and
      Actuator, 15: 220-230.

https://doi.org/10.1016/j.sna.2009.02.032

[22]   Roach, D. 2009. Real time crack detection using mountable comparative vacuum monitoring sensors, Smart Structures and Systems, 5(4): 317-328.

https://doi.org/10.12989/sss.2009.5.4.317

[23]   Ryu, C., Lee, J., Kim, C., and Hong, C. 2008. Buckling behavior monitoring of a composite wing box using multiplexed and multi-channel built in fiber Bragg grating strain sensors, NDT & E, 41: 534-543.

https://doi.org/10.1016/j.ndteint.2008.05.001

[24]   Salas, K.I. and Cesnik, C.E.S. 2009. CLoVER: An Alternative concept for damage interrogation in structural health monitoring systems, The Aeronautical Journal, 113(1144): 339- 357.

https://doi.org/10.1017/S000192400000302X

[25]   Scruby, C.B. and Drain, L.E. 1990. Laserultrasonics—techniques and applications. England: IOP Publishing

[26]   Sonatest Inc. 2013. Sitescan D+ Series, Sonatest Limited, Part No. 147359, Issue 2, Product Brochure retrieved on 28 August 2013 from http://www.sonatest.com/products/range/ transducers/probes/single/.

[27]   Staszewski, W.J, Mahzan, S. and Traynor, R. 2009. Health monitoring of aerospace composite structures-active and passive approach, Composite Science and Technology, 69: 1678-1685.

https://doi.org/10.1016/j.compscitech.2008.09.034

[28]   Valdes S.H.D. and Soutis C. 1999. Delamination detection in composite laminates from variation of their modal characteristics", Journal of Sound and Vibration, 1: 1-9.

https://doi.org/10.1006/jsvi.1999.2403

[29]   Verijenko, B. and Verijenko, V. 2005. A new structural health monitoring for composite laminates, Composite Structures, 21: 315-319.

https://doi.org/10.1016/j.compstruct.2005.09.024

[30]   Wang, S., Kovalik, D.P. and Ching, D.D.L. 2004 Self sensing attained in carbon fiber polymer matrix structural composites by using the interlaminar interface as a sensor, Smart Material Structure, 13: 570-592.

https://doi.org/10.1088/0964-1726/13/3/017

[31]   White, C., Herszberg, I. and Mouritz, A.P. 2009. Structural Consequences of Sensor Cavities In Scarf Repairs, Materials Forum. 33: 427-434

[32]   Whittingham, B., Li, H.C.H., Herszberg, I. and Chiu, W.K. 2006. A disbond detection in adhesively bonded composite structures using vibration signatures Composite Structures, 75: 351–363.

https://doi.org/10.1016/j.compstruct.2006.04.055

[33]   Zhang, H., Schulz M.J. and Feruson, F. 2002. Structural health monitoring using transmittance functions, Mechanical Systems and Signal processing, 2: 357-378.

Chapter 4

Damage identification for composite panels using PZT sensor

Nisreen N. Ali Al-Adnani[1,*], F. Mustapha[2]

Department of Civil Engineering, Universiti Putra Malaysia, 43400 Serdang, Selangor, Malaysia

Department of Aerospace Engineering, Universiti Putra Malaysia, 43400 Serdang, Selangor, Malaysia,

Keywords

Structural Health Monitoring (SHM), Damage Identification, Twill Weave Carbon Fibre, Composites, Smart Sensor, Root Mean Square Deviation

Abstract

Real-time monitoring of structural integrity is an important challenge. This article presents the results of damage detection in real time for two materials: Al 6061-T6 and twill weave carbon fibre-reinforced epoxy composite. The natural frequency as a global dynamic technique was adopted and the structure was evaluated based on the change in the natural frequency. A square thin plate with simply supported edges was investigated under the effect of sinusoidal signal which was generated via mechanical vibration exciter to carry out the natural frequency of the panel. A smart sensor (piezoelectric ceramic lead zirconate titanate) bonded to the surface of the composite panel was used to capture the signals. Experiments demonstrate the effect of change in crack depth and the response of these panels. The results were measured via monitoring technique and evaluated using root mean square deviation index as statistical analysis.

Contents

1. Introduction

Damage detection existences in Civil, aerospace and mechanical structures can enhance the safety, security, and extend the structural service life, and reduce the maintenance costs. Early detection of the damage or structural degradation prior to local failure can prevent a catastrophic collapse of those structures. Typical damage in these infrastructures might be due to the development of cracks. Recently, extensive research works in civil, aerospace applications have been extended by using fiber-reinforced plastic composite materials. These materials contain strong and continuous fibers bound together by a continuous matrix of polymer resin. Structural vibration control along with smart materials is gradually being used for flexible structures and it has attained remarkable progress. Vibration control is critical to the development of advanced lightweight and structures such as: helicopter, wind turbine blades, aircraft wing and flexible space structure such as a solar array, light weight trusses and space station (Schulz et al., 1999). PZT sensor worked as a link between the non-electrical setting and data processing electric schemes (Waanders, 1991). Novel smart sensors and actuators, such as Piezoelectric Ceramic Lead Zirconate Titanate (PZT) transducers have been identified as the method of Structural Health Monitoring (SHM) technology development and it is widely used for monitoring requests. For SHM approaches, the data reduction technique based on principal component analysis was applied by Aris et al., (2014). Two carbon fiber-reinforced plastic panels were subjected to damage and repair coinciding with typical aircraft repair procedures. Via placing smart PZT sensors type (APC-850), the raw data in a Lamb waveform were captured at 100 mm across the damaged and repaired structures. The results presented the talented accuracy and repeatability of the data and it was possible to distinguish the circumstances of undamaged, damaged, and repaired cases. In addition, Aris et al., (2015) studied two experimental procedures, normal, damaged and repaired conditions in aircraft panels using PZT sensor. Carbon Fiber-Reinforced Plastic (CFRP) was the first case. A function generator and the oscilloscope were used. The result showed the ability of the sensors to sense structural integrity in two cases as normal and repaired samples. Where, Wu et al., (2009) developed a hybrid PZT/FBG system based on composite laminate plates to detect the damage. Piezoelectric actuators were used to control the input excitation to the structure while fiber optic sensors were used to capture an aluminum plate response. The results demonstrated the viability of identifying the simulated damage on an aluminum plate by the proposed hybrid system. Moreover, for damage detection and localization in a composite plate (Zumpano and Meo, 2008) presented a novel Transient Non-Linear Elastic Wave Spectroscopy (TNEWS). Based on time-frequency and coherence function and via two different pulse excitation amplitudes, the TNEWS analysed the un-connections between two structural dynamic responses. The developed method recognized and reflected robustness and a perfect way to monitor the non-linear elastic wave propagation performance. However, the bolt load loss in hybrid metal composite connections was investigated by Caccese et al., (2004a) and Caccese et al., (2004b) via frequency domain techniques. A composite plate fabricated using E glass with a 6.35 mm in thickness and bolted to a steel frame. Controlled vibration input was provided via a PZT actuator bonded to the center of the composite panel. Damage index approach was used to evaluate the effect of load changing. The results improved and the transmittance function approach had the most capacity.

Natural frequency as a basic vibration parameter has advantages: it is easy to measure, is less influenced by environmental noise; besides it is considered as the smartest for a fast and global method that can be measured at one single position in the structure (Srinivas et al., 2009). Commonly, in structural assessment, the natural frequency is used as an investigative parameter for vibration monitoring and the structural circumstance can be monitored via analyzing the periodical frequency measurement. In addition, if there are cracks with similar crack length in two different locations it may affect the same quantity of frequency changes, and the changes in the natural frequency may not be acceptable for selecting the identification of the location of structural damage (Salawu, 1997). Any damage that occurs can affect the structural condition and change the signal (Park and Inman, 2007). And, severity in structure due to damage can be measured directly when the natural frequency in the structure changes (Kim et al., 2003). A frequency response method in SHM system has many advantages such as: it is cheap, light in weight, and provides understanding as to the overall condition of the system (Kessler et al., 2002). Some measurements are inflexible in the time domain but it is easy in the frequency domain. Measurement of the structure's natural frequency can provide a non-destructive testing technique (Cawley and Sarsentis, 1988). The structure integrity can be monitored via periodical frequency measurements based on the sensitivity of natural frequency as a good indicator for damage detection. Also, Cawley and Sarsentis, (1988) considered the natural frequency as a faster technique than other methods such as ultrasonic process which requests a transducer to scan over all the examination spaces. In addition, this type of test is smart and it is possible and reliable to define the properties of a whole structure via a single point measurement. The spectrum is the fundamental measurement and the spectrum's magnitude is displayed to represent the complete signal amplitude in each frequency set (SRS, 2015). The distribution of the frequency differences may be described either by a measure or by a statistical occupation. It is significant to use the power spectrum in statistical analysis of signal processing (Spectral Density, 2015). Coherence is used to examine the record sets. It is usually used to evaluate the power transfer and the joining of input and output. The accuracy of the signal data can be reflected via the coherence curve. The damage identification is one objective of this study in order to predict the integrity of the structure. Therefore, similar procedures were performed in this study by using the data of the two materials for statistical analysis based on Root Mean Square Deviation (RMSD) index.

2. Materials and instrumentation

Two materials were used in this investigation, mechanical vibration exciter, and NI DAQ and LabVIEW software.

a) Aluminium Alloy Type 6061-T6

Structural prototype to simulate a three- storeyed building was constructed by using aluminum alloy type 6061-T6 with dimensions 300 mm x 300 mm x 3 mm. The floor was considered as a reference to compare with GFW/ Epoxy composite laminate panel. The floor's design was capable of connecting with the columns. The corner was cut and removed and squared in shape with dimensions of 40 mm x40 mm as shown in Figure 1.

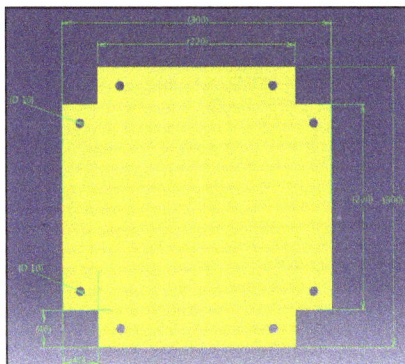

Figurer 1. Dimensions of Floor Panel.

b) Composite Panels

The specification of glass fiber, which is used in this research to fabricate the composite panels, is glass fibre type Plain Weave 300 g/m^2 with warp and weft yarns arranged parallel and flat. Composites were fabricated using hand lay-up and vacuum bagging method to produce the laminate composite planes (GFW/ Epoxy) and in the same thickness of Al 6061-T6 (3 mm).

c) Frame Structure's Design

In this study, the prototype was constructed using aluminium frame fixed on the mechanical vibration exciter (shaker table) in order to test the selected materials under the effect of vibration. The three-storeyed frame experimental setup was used for the two types of materials. The experimental setup included the frame with composite panels of three floors. The experiment was designed to realise and approve a SHM technique for a composite application. Figure 2 illustrates the frame design in detail.

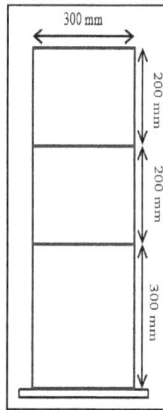

Figure 2. Aluminium Frame Setup (Front View).

The first case studies were carried out to check the feasibility of undamaged detection, it was considered as a reverence to compare with the other damage cases. Three types of damage were presented as: 10 mm, 15 mm or 20 mm crack's depth as shown in Figure 3 (a, b and c), respectively.

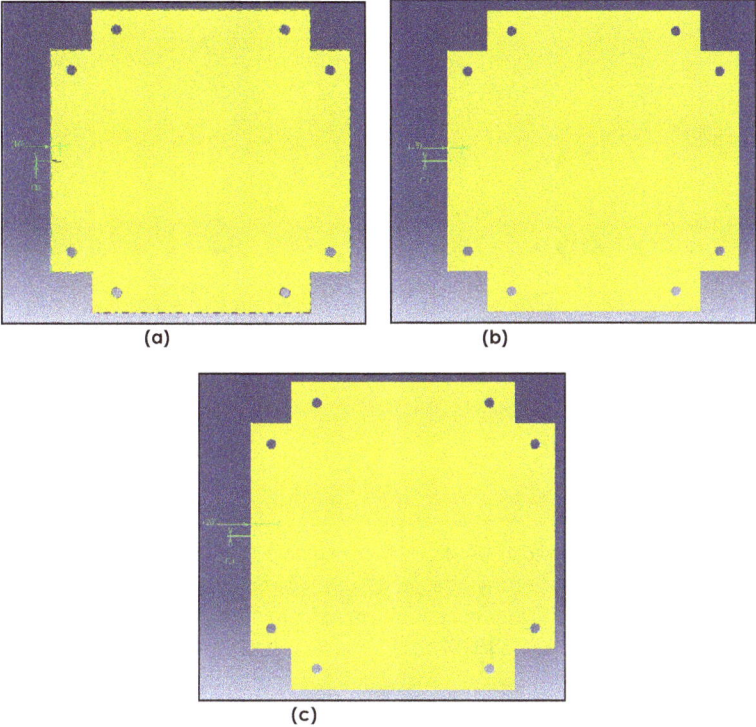

Figure 3. Square Panel with Crack Lengths of: (a) 10 mm, (b) 15 mm, and (c) 20 mm.

In Figure 4, three panels for two materials were cut according to the aluminium frame's design and prepared to be fixed as three floors. These panels were changed and fixed individually based on the same material for all the three floors.

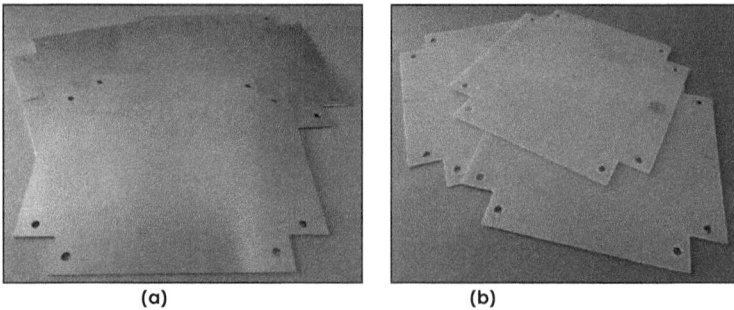

(a)   (b)

Figure 4. Three Panels of Two Materials: (a) Al 6061-T6, and (b) GFW/ Epoxy Composite.

d) Data Acquisition (DAQ)

PZT materials were utilised as a powerful and innovative tool for local damage detection of various structures. APC International supplied PZT 850 with these characterisations which also mentioned approval by the APC Company: high in charge sensitivity, link, and density with fine particle structure, as well as it has a large displacement, noise-free frequency response, and a higher operating temperature. Data Acquisition (DAQ) device involves NI-9234 module and NI USB-9162 high-speed carrier. National Instrument and LabVIEW software is a graphical programming environment for developing refined measurement, test, and control systems (LabVIEWTM SinalExpress, 2012). In this investigation, PZT 850 sensor type disk is 10 mm in diameter and 1 mm in thickness. This PZT was connected by wire to the NI USB-9234 DAQ, while the NI 9234-USB was linked to the laptop. LabVIEW SignalExpress software was installed and employed to analyse the collected data. PZT sensor was bonded on the surface at mid-span edge of 3rd floor panel to monitor, capture, log, and to analyse real-time data for the performance of undamaged structures and three cases with different types of damage in structures. The location of crack was selected as a fixed entity for all the case studies while the depth was considered as varied. In this investigation, damage detection and identification adopted the frequency domain which was filtered to ensure that it was free of noise effects. The analysis was performed by using the natural frequencies and power spectrum. The variation of the scattered signal can be described via the power spectrum within a sequence time, and present the power of that signal at each frequency which may be decomposed and distributed. The performance of the power spectrum on the complete signal can be spread into short parts.

3. Root mean square deviation (RMSD) index for damage identification

One of the dynamic properties like natural frequency is related to the structural properties and it is able to detect any defect in the structure. To date, several damage metrics are active to compare and measure the existence of damage. The distribution of the frequency differences may be described either by a measure or by a statistical occupation. One of the statistical analyses is the Root Mean Square Deviation (RMSD). The damage can be quantified via root mean square deviation of the natural frequency when the crack is propagated and then the damage index can be released. The mean is a helpful function to organise and analyse large sets of data. Mostly, RMSD as a non-parametric measurement was used to quantify damage (Min et al., 2011; Neto et al., 2011; Annamdas et al., 2010; Panigrahi et al., 2010; Yang et al., 2010; Baptista and Filho, 2009; Park et al., 2006; Giurgiutiu and Zagrai, 2005; Rutherford et al., 2004; Caccese et al., 2004; Naidu and Soh, 2004; and Raju, 1997). RMSD is defined as per the following equations:

$$\text{RMSD} = \sqrt{\sum_{i=1}^{n} \frac{(Z_{1,i} - Z_{2,i})^2}{n}} \tag{1}$$

where:

Z1 and Z2: are the real parts of the baseline measurement and measurement used for comparison respectively,

i: is the frequency interval, and

n: is the total number of frequency points used in the comparison (Yang et al., 2009).

Alternatively, the RMSD is sometimes scaled by the baseline values instead of the number of points as defined in the following:

$$\text{RMSD} = \sqrt{\sum_{i=1}^{n} \left\{ \frac{(Z_{1,i} - Z_{2,i})^2}{Z_{1,i}^2} \right\}} \tag{2}$$

The results from these equations cannot be considered as a result of damage. The mean of each measurement can be subtracted from the measurement to produce the damage metric as per the following equation, and the form of the RMSD damage metric then becomes:

$$\text{RMSD} = \sqrt{\sum_{i=1}^{n} \left\{ \frac{\left((Z_{1,i} - Z_{1}^{-}) - (Z_{2,i} - Z_{2}^{-})\right)^2}{n} \right\}} \tag{3}$$

where:

Z^-: indicates the mean of the measurement (Peairs, 2006 and Raju, 1997).

4.   Experimental setup

The purpose of the experiment was to evaluate SHM technique using PZT sensor and to detect the crack damage in Al 6061-T6 and GFW/ Epoxy composite structures in real time. The acquired output signals, frequency domain values can be obtained to proceed further with the statistical analysis. The data were collected from each structure as undamaged (healthy) and damaged. Through the experiment, an excitation was provided via mechanical vibration exciter. The excitation frequency was chosen and controlled as a same value for all specimens' test sets. The natural frequency responses were measured for all panels due to the effect of that vibration. The processing of the signals was accomplished using LabVIEW SignalExpress software to compare the experimental results for validation and identification. The experimental setup is demonstrated by Nisreen et al., 2015.In each test, the PZT sensor was bonded in the mid-span of the 3rd floor edge of the host structure and in the same portion to prevent any difference with the baseline measurements otherwise every sensor would measure a different dynamic response. The boundary condition of the floor's panel was four edges and it was simply supported. In NI USB-9234 DAQ, a physical channel is a pin setting and connected to the sensor to measure an analogue signal.

The first test for the 3rd floor of the structure (prototype) in undamaged (healthy) case was done for Al 6061-T6 (as a reference case). PZT was bonded on the surface of the panels in the consulted position to capture the frequency. The first case was undamaged for the two materials and considered as references to compare the damaged cases. Once, a significant result from the healthy specimens was obtained then it was moved to the second step of the experiment. Experimentally, the cracks were formed with an electric saw in the mid-span of the panels to simulate the damage with three different lengths: 10 mm, 15 mm and 20 mm. These damages were created in the two materials. In each case, PZT was bonded on the internal end of the crack. PZT sensor was applied to the structure to capture the vibration signals. Any change which occurred in the state of the structure was established as a deviation in these signatures and considered as SHM and Non-Destructive Evaluation (NDE) technique (Peairs, 2006). The experimental setup for the Al 6061-T6 and GFW/ Epoxy panels is presented in Figure 5 (a and b), respectively.

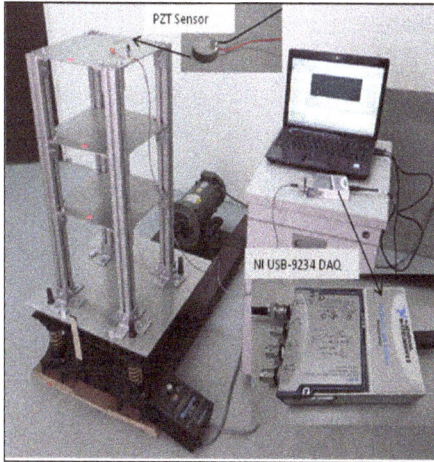

(a)

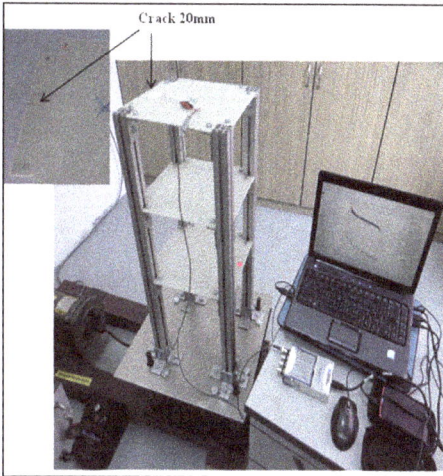

(b)

Figure 5. Experimental Setup: (a) Al 6061-T6, and (b) GFW/ Epoxy Composite.

5. Results and discussions

One of the dynamic parameters is the natural frequency and the fact that any change due to damage or deterioration in any structure can affect the natural frequency. Monitoring these changes in frequencies can reveal the structure's integrity if it is undamaged or if any damage may occur. Based on that, the four case studies for two materials were evaluated to identify the ability of these materials and their resistance against the vibration effects. The results were measured and evaluated using SHM technique including: NI (DAQ USB-9234) device and LabVIEW SignalExpress software. The changes over the signals period were measured with the amplitude which was adopted as a function of magnitude to present the difference between the extreme values for the single variables. In general, the signals were filtered and in acceleration function and time domain. Power spectrum was presented vs frequency. The Coherence was demonstrated as amplitude vs frequency, starting from 0 to 13000 Hz as the signal's test limitation. In the undamaged case, the signal result of the natural frequency was considered as the baseline measurement. In the undamaged case, four sets of data were obtained to present the filtered signals, power spectrum, zoom power spectrum and coherence for two materials. The same procedures were carried out for the changes in the conductance signature as observed for the three damaged cases. The signal power which passed through the filter was measured to determine the signal strength in certain frequency bands. For Al 6061-T6 and GFW/ Epoxy, the harmonic excitation signals have almost the same model of frequency with different peak values. The acceleration values varied between +0.01 and -0.015 for both Al 6061-T6 and GFW/Epoxy. The filtered signal was recorded as a function of acceleration versus the time as shown in Figure 6 (a, b, c and d).

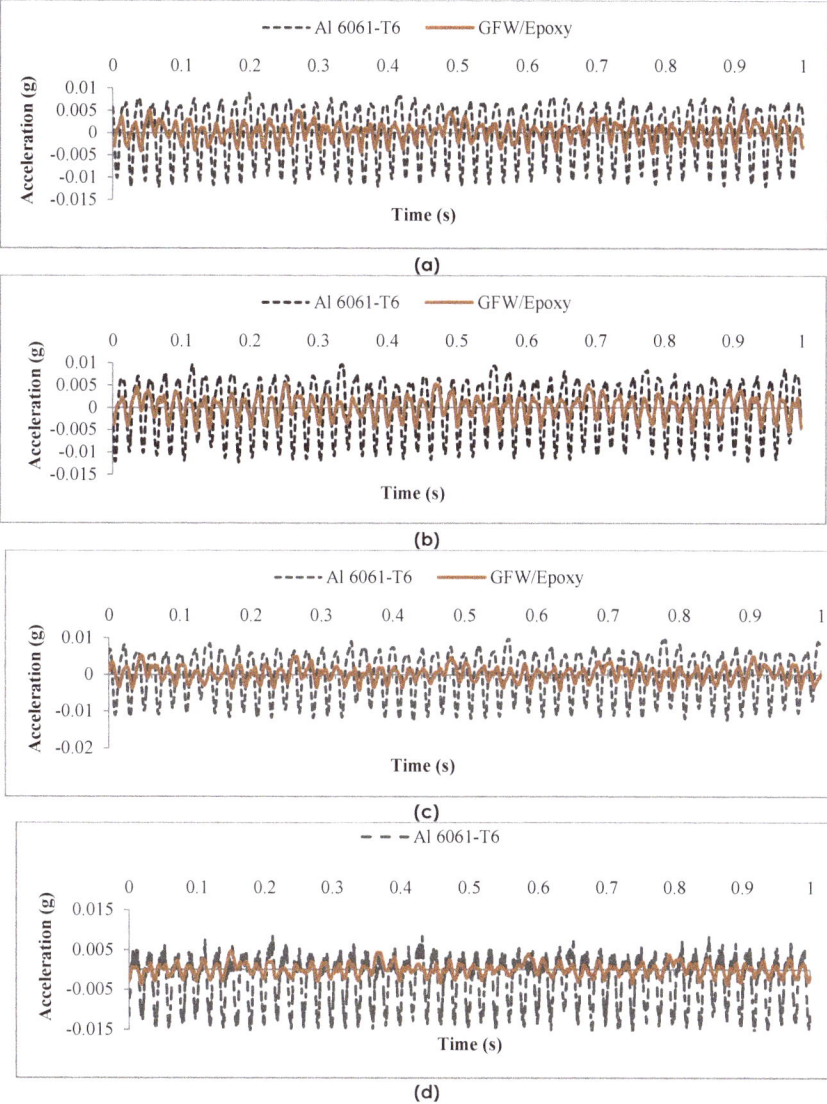

Figure 6. Filtered Signals of Al 6061-T6 and GFW/ Epoxy: (a) Undamaged, (b) Crack 10 mm, (c) Crack 15 mm, and (d) Crack 20 mm.

The coherence was recorded as 1, as evidence of the collected data's accuracy in Al 6061-T6 and GFW/Epoxy. The coherence patterns are the same in all the four case studies (one undamaged and three damaged), Figure 7 shows the coherence pattern in LabView software analysis.

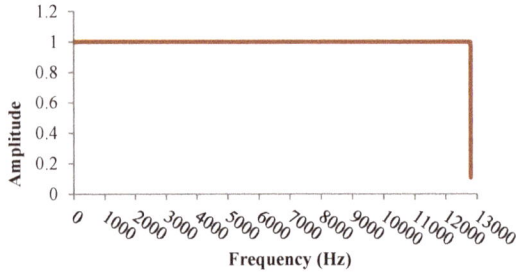

Figure 7. Coherence Result.

Comparison of the signals results presented the various peak values of frequencies which were affected by the material's properties. The Zoom Spectrum Interval was selected for all the data to compare the variation in signals. The metal material's response like Al 6061-T6 was considered in this investigation as the baseline measurement. These undamaged signals were considered as a baseline measurement to compare with the other cases (three cracked damage). There were clear differences between the natural frequency patterns that were properties of these materials and the crack effect on the frequency patterns. The frequency shifted slightly due to the vibration effect on the panels with occurred cracks.Through the peak amplitude finding in the complete response, the curves of peak amplitude response were formed and plotted. Comparision between the zoom spectrum of two investigated materials in the cases: undamaged, crack 10 mm, crack 15 mm, and crack 20 mm are shown in Figure 8 (a and b), respectively.

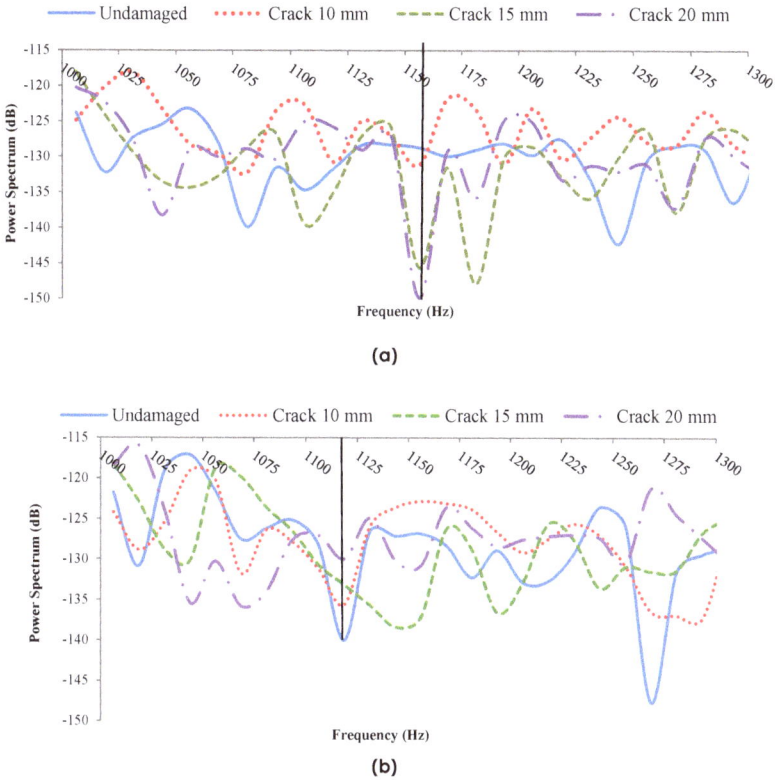

Figure 8.   Zoom Spectrum: (a) Al 6061-T6, and (b) CFW/Epoxy Composite.

Zoom power spectrum values were considered within the frequency limitation between 1000 Hz and 1300 Hz to compare Al 6061-T6 and GFW/ Epoxy responses. The nature of the persuaded damages was practically small. Commonly, the appearance of new peaks in the signals is due to the effect of damage which is observed by lateral and vertical shifts of the peaks. These peaks are very small and cannot reflect clearly the vibration responses of the panels. There are small random variations along the curves. However, the variations are relatively small. After the damage is induced, a significant change occurs in the signature pattern of the impedance curve over the entire frequency range. This is because the damage causes changes in stiffness resulting in changes in mechanical impedance of panel. However, the drop in peak amplitude curves between the undamaged panel and the crack damaged in the panels was clear. The impedance signatures significantly changed the response frequencies and tende to shift left which induced the damage. Based on the structure integrity state, the decrease in peak amplitude was observed as a drop in peak amplitude in the damaged cases in Al 606-T6 and GFW/ Epoxy. Non-homogenous materials have unique responses. The composite panels were non-homogenous materials therefore the natural frequency varied. At Frequency 1120 Hz, only slight unequal changes were found in the peak amplitude for GFW/ Epoxy panel with Crack 20 mm. The RMSD's results of Al 6061-T6 and GFW/ Epoxy in four cases were plotted as undamaged, crack 10 mm, crack 15 mm and crack 20 mm. Al 6061-T6 was considered as the reference to compare RMSD values with CFW/Epoxy. In order to quantify the change in frequency due to damage, RMDS index is utilized. It is calculated by following Equation 3. All calculations in this paper have been made using only the real part of the impedance. The final comparison of RMSD % between collected healthy and damaged results was evidence that when the damage increased, the RMSD was also increased. RMSD values of Al 6061-T6 and GFW/ Epoxy were very close. It is concluded from the results that the GFW/ Epoxy has the ability to resist the crack. Additionally, it means that the GFW/ Epoxy has a high performance to resist the frequency affected by vibration, and the results are presented in Figure 9.

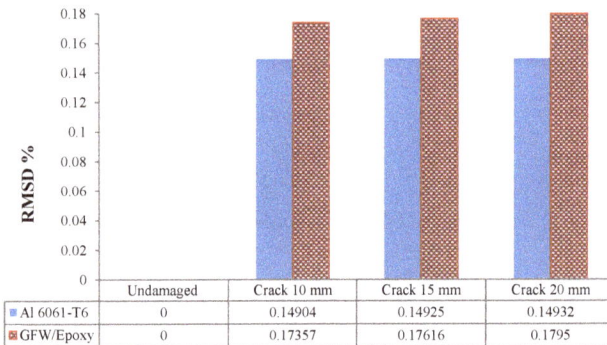

	Undamaged	Crack 10 mm	Crack 15 mm	Crack 20 mm
■ Al 6061-T6	0	0.14904	0.14925	0.14932
⊠ GFW/Epoxy	0	0.17357	0.17616	0.1795

Figure 9. RMSD% Results of Al 6061-T6 and GFW/ Epoxy.

6.   Conclusion

The effort summarized in this study focuses on real time damage diagnosis based on SHM method when PZT sensor was employed in Al 6061-T6 and CFW/Epoxy laminated composite panels. Frequently, mid-span is critical to the function of the structure and its failure may incur high repair costs, or may risk survival. From the experimental results of the mechanical vibration exciter test, the damage indices were obtained by the proposed SHM technique. SHM monitoring method can monitor and predict the health status of the tested panels through the damage index matrix. Four cases were investigated: undamaged as a 1st case and considered as a reference, in addition to three cracks created with length of 10 mm, 15 mm and 20 mm to simulate 2nd, 3rd and 4th cases, respectively. The significance of vibration effects on selected specimens provides the effectiveness of the SHM system for damage identification and for composite panels. PZT sensor has the potential to be applied in health monitoring of composite structures. Statistical analysis is used to compare the results. A good performance is observed for GFW/Epoxy structure. The GFW/Epoxy as talent material is able to struggle against progression and prevent increase in the crack's depth. In general, the natural frequency results are decreased as the structure's stiffness is decreased and weakened due to crack damage. RMSD is an acceptable statistical analysis to evaluate structural integrity. Qualitative health monitoring technique is numerous: condensed maintenance costs, improved reliability and safety and increased structural lifetime.

Acknowledgment

The authors gratefully acknowledge the financial support provided by Universiti Putra Malaysia under Research University Grant Scheme no. [RUGS- 9348100].

References

[1]   Annamdas, V. G. M., Yang, Y., and Soh, C. K. (2010). Impedance based concrete monitoring using embedded PZT sensors. International journal of civil and structural engineering. 1(3): 414-424.

[2]   Aris, K. M., Mustapha, F., Sapuan, S., and Majid, D. A Structural Health Monitoring of a Pitch Catch Active Sensing of PZT Sensors on CFRP Panels: A Preliminary Approach. DOI: 10.5772/48097. Book, Chapter 1. Retrieved 23 Jan 2015 from:
https://doi.org/10.5772/48097

[3]http://www.stamplive.com/apu.php?n=&zoneid=8191&cb=1725730654&popunder=1 &direct=1.

[4]   Aris, K. D. M., Mustapha, F., Salit, M. S., and Majid, D. L. A. A. (2014). Condition Structural Index using Principal Component Analysis for undamaged, damage and repair conditions of carbon fiber-reinforced plastic laminate. Journal of Intelligent Material Systems and Structures, 25(5), 575-584.

https://doi.org/10.1177/1045389X13494932

[5]   Baptista, F. G. and Filho, J. V. (2009). A new impedance measurement system for PZT-based structural health monitoring. Instrumentation and Measurement, IEEE Transactions on, 58(10), 3602-3608.

https://doi.org/10.1109/TIM.2009.2018693

[6]   Caccese, V., Mewer, R., and Vel, S. S. (2004a). Detection of bolt load loss in hybrid composite/metal bolted connections. Engineering Structures, 26(7), 895-906.

https://doi.org/10.1016/j.engstruct.2004.02.008

[7]   Caccese, V., Richard Mewer, a., and Vel, S. S. (2004b). Detection of Bolt Load Loss Using Frequency Domain Techniques, October 24-27, Bar Harbor, Maine, USA. Proceeding of the 15th International Conference on Adaptive Structures and Technologies.

[8]   Cawley P. and Sarsentis N. (1988). A Quick Method for the Measurement of Structural Damping. Mechanical System and Signal Processing, 2(1): 39-47.

https://doi.org/10.1016/0888-3270(88)90050-7

[9]   Giurgiutiu, V., and Zagrai, A. (2005). Damage detection in thin plates and aerospace structures with the electro-mechanical impedance method. Structural Health Monitoring. 4(2): 99-118.

https://doi.org/10.1177/1475921705049752

[10]   Kessler, S. S., Spearing, S. M., Atalla, M. J., Cesnik, C. E., & Soutis, C. (2002). Damage detection in composite materials using frequency response methods. Composites Part B: Engineering, 33(1), 87-95.

https://doi.org/10.1016/S1359-8368(01)00050-6

[11]   Kim, J.-T., Ryu, Y.-S., Cho, H.-M., and Stubbs, N. (2003). Damage identification in beam-type structures: frequency-based method vs mode-shape-based method. Engineering Structures, 25(1), 57-67.

https://doi.org/10.1016/S0141-0296(02)00118-9

[12]   LabVIEWTM SinalExpress, Getting started with LabVIEW SignalExpress, National Instrument June 2012, Manual.

[13]   Min, J., Shim, H., and Yun, C.-B. Electromechanical Impedance-based Damage Identification Using Multiple Piezoelectric Sensors. The 6th International

Workshop on Advaced Smart Materials and Smart Structures Technology (ANCRiSST 2011) July 25-26, 2011, Dalian. China.

[14]   Naidu, A., and Soh, C. (2004). Damage severity and propagation characterization with admittance signatures of piezo transducers. Smart Materials and Structures, 13(2), 393.

https://doi.org/10.1088/0964-1726/13/2/018

[15]   Neto, R. M. F., Steffen, V., Rade, D. A., Gallo, C. A., and Palomino, L. V. (2011). A low-cost electromechanical impedance-based SHM architecture for multiplexed piezoceramic actuators. Structural Health Monitoring, 10(4), 391-402.

https://doi.org/10.1177/1475921710379518

[16]   Nisreen N. Ali, F. M., S. M. Sapuan,R. S. M. Rashid (2015). An Approach and Experimental Technique for Damage Detection of Composite Panels Using PZT Sensor. International Journal of Civil and Structural Engineering Research 3(1), 29-38.

[17]   Park, G., and Inman, D. J. 2007. Structural health monitoring using piezoelectric impedance measurements. Philosophical Transactions of the Royal Society A: Mathematical, Physical and Engineering Sciences. 365(1851): 373-392.

https://doi.org/10.1098/rsta.2006.1934

[18]   Park, G., Farrar, C. R., di Scalea, F. L., and Coccia, S. (2006). Performance assessment and validation of piezoelectric active-sensors in structural health monitoring. Smart Materials and Structures. 15(6): 1673.-1683.

[19]   Panigrahi, R., Bhalla, S., and Gupta, A. (2010). A Low-Cost Variant of Electro-Mechanical Impedance (EMI) Technique For Structural Health Monitoring. Experimental Techniques, 34(2), 25-29.

https://doi.org/10.1111/j.1747-1567.2009.00524.x

[20]   Peairs, D. M. (2006). High frequency modeling and experimental analysis for implementation of impedance-based structural health monitoring. PhD. Dissertation, Virginia Polytechnic Institute and State University.

[21]   Raju, V., (1997). Implementing Impedance-based Health Monitoring, Master's thesis, Virginia Polytechnic Institute and State University, Blacksburg, Virginia.

[22]   Rutherford, A. C., Park, G., Sohn, H., and Farrar, C. R. (2004). The Use of Electrical Impedance Moments for Structural Health Monitoring. Paper presented at the Proceedings of the 22nd IMAC.

[23]   Salawu, O. S. 1997. Detection of structural damage through changes in frequency: a review. Engineering Structures, 19(9): 718-723.

https://doi.org/10.1016/S0141-0296(96)00149-6

[24]   Schulz, M., Pai, P., & Inman, D. (1999). Health monitoring and active control of composite structures using piezoceramic patches. Composites Part B: Engineering, 30(7), 713-725.

https://doi.org/10.1016/S1359-8368(99)00034-7

[25]   Spectral Density. Retrieved 23 Jan 2015 from: http://en.wikipedia.org/wiki/Spectral_density.

[26]   Srinivas, V., Sasmal, S., & Ramanjaneyulu, K. (2009). Studies on methodological developments in structural damage identification. Structural Durability and Health Monitoring, 5(2), 133-160.

[27]   SRS, Stanford Research System. Retrieved 23 Jan 2015 from: www.thinkSRS.com.

[28]   Waanders, J. W. (1991). Piezoelectric Ceramics, Proprties and applications, Philips Components, EINDHOVEN, The Netherlands. First Edition April 1991.

[29]   Wu, Z., Qing, X. P., and Chang, F.-K. (2009). Damage detection for composite laminate plates with a distributed hybrid PZT/FBG sensor network. Journal of Intelligent Material Systems and Structures. (9 pp.).

[30]   Yang, Y., Divsholi, B. S., and Soh, C. K. (2010). A reusable PZT transducer for monitoring initial hydration and structural health of concrete. Sensors, 10(5), 5193-5208.

https://doi.org/10.3390/s100505193

[31]   Yang, Y., Liu, H., Annamdas, V. G. M., and Soh, C. K. (2009). Monitoring damage propagation using PZT impedance transducers. Smart Materials and Structures, 18(4), 045003, (9 pp.).

https://doi.org/10.1088/0964-1726/18/4/045003

[32]   Zumpano, G., and Meo, M. (2008). Damage localization using transient non-linear elastic wave spectroscopy on composite structures. International Journal of Non-Linear Mechanics, 43(3), 217-230.

https://doi.org/10.1016/j.ijnonlinmec.2007.12.012

Chapter 5

The bonded macro fiber composite (MFC) and woven kenaf effect analyses on the micro energy harvester performance of kenaf plate using modal testing and Taguchi method

A. Hamdan[1], F. Mustapha[2]

[1]Department of Aerospace Engineering, Universiti Putra Malaysia, 43400 Serdang, Selangor, Malaysia

[2]Aerospace Manufacturing Research Centre (AMRC), Level 7, Tower Block, Faculty of Engineering, Universiti Putra Malaysia, 43400 Serdang, Selangor, Malaysia

Keywords

Modal Testing, Taguchi Method, Energy Harvester, Kenaf Fiber Composite, Macro Fiber Composite

Abstract

The number of wind energy applications will continue to increase as fossil fuel reservoir keeps decreasing. More researches are recently conducted to employ a green material concept for turbine blades. The usage of natural fiber reinforced composite, especially kenaf fiber, in the fabrication of wind turbines needs to be given due attention. Woven and unwoven kenaf fiber is employed to fabricate composite plates which replicate the simple turbine blade model. In addition, Macro Fiber Composite (MFC) is attached to the kenaf plates for structural health monitoring and micro energy harvester purposes. The MFC used is attached with two techniques which are surface bonded and embedding into the plate. The modal testing analysis and Taguchi method is employed to investigate the effects of attached MFC technique. Bonded technique is suggested as the most influenced factor in micro energy harvesting at the vibration range of 20 to 60 Hz. Furthermore, the kenaf woven type, the distance from structure neutral axis, the stiffness of structure, the excitation vibration and the neutral frequency of a structure are highlighted as the factors influencing the performance of micro energy harvester.

Contents

## 1.	Introduction

The study on wind resource density has already been conducted by several researchers in Malaysia. A ten years study from 1982 to 1991 at 10 different areas in Malaysia suggested that Mersing and Kuala Terengganu have a wind power potential with a mean power density of 85.61 W/m^2 and 32.50 W/m^2 respectively (Sopian, Othman et al. 1995; Hamdan, Mustapha et al. 2014). Moreover, the annual offshore wind speed in Malaysia is around 1.2–4.1m/s with the east coast of Peninsular Malaysia recording the highest in annual vector resultant wind speed of 4.1m/s (Masseran, A.M.Razali et al. 2012). The stability of wind speed in peninsular Malaysia is known to be quite good (Masseran, Razali et al. 2012). This shows that the application of wind energy harvesters is relevant especially at places with sufficient wind resources. Furthermore, the rural areas of east Malaysia have already utilized hybrid solar and wind energy generators (Hamdan, Mustapha et al. 2014).

The application of vertical axis wind turbine (VAWT) in wind energy generator has created advantages in several aspects such as free wind direction oriented, the ability to be towerless and huge power density per square meter (Hamdan, Mustapha et al. 2014). However, material for turbine blades has raised concern. The application of synthetic fibers such as glass fiber and carbon fiber in turbine blades have promoted several disadvantages such as risk of inhalation during the fabrication process, renewability, biodegradability and recyclability issues. The usage of biocomposite fiber to replace synthetic fiber as a reinforcement in Fiber-Reinforced Plastics (FRP) is beginning to widespread (Shalwan and Yousif 2013). Investigation in biocomposites have lead to several types of natural fibers such as flax (Oksman, Skrifvars et al. 2003; Stuart, Liu et al. 2006), bamboo (Lee and Wang 2006), pineapple (Liu, Misra et al. 2005), jute (Plackett, Andersen et al. 2003; Wambua, Ivens et al. 2003) and kenaf (Nishino, Hirao et al. 2003; Wambua, Ivens et al. 2003; Cao, Goda et al. 2007; Ochi 2008). The mechanical properties in natural fiber may differ due to several factors such as fiber morphology, structure density, cell wall thickness, woven or nonwoven (Yahaya, Sapuan et al. 2014), length and diameter of the structure (Faruk, Bledzkia et al. 2012). Kenaf fiber is one of the natural fiber which is extensively researched and has given significant findings for future benefits (Zampaloni, Pourboghrat et al. 2007; Davoodi, Sapuan et al. 2010; Akil, Omar et al. 2011; Faruk, Bledzkia et al. 2012; Yahaya, Sapuan et al. 2014).

Several researches on kenaf polymer have already been conducted such as analysis studies on the effect of chemical treated on kenaf (Yousif, Shalwan et al. 2012 ; Yahaya, Sapuan et al. 2015), fiber fraction (Nishino, Hirao et al. 2003; Yahaya, Sapuan et al. 2014), woven orientation (Azrin Hani, Chan et al. 2013; Hani, Ahmad et al. 2013; Yahaya, Sapuan et al. 2015), hybrid effect (Davoodi, Sapuan et al. 2010) and resin application (Wambua, Ivens et al. 2003; Aziz and Ansell 2004; Ochi 2008; Shibata, Cao et al. 2008). A few elements are highlighted in the study that influence the mechanical properties of natural fiber such as the volume fraction, the interfacial adhesion of the fibre with the matrix and orientation of the fibre and length of the fiber (Yousif, Shalwan et al. 2012 ; Reza, Jamaludin et al. 2014). These reports are summarized in Table 1. R1, R2 and R3 presented in Table 1 denoted the sample that will be compared in term of mechanical properties later on.

Table 1 Mechanical properties of kenaf composite from previous researchers

Sample	Reinforce form	Resin	Maximum Stress (Mpa)	Flexural Modulus (Gpa)	Tensile strength (MPa)	Tensile Modulus (GPa)
(R1)Woven Kenaf (Untreated NaOH) (Yahaya, Sapuan et al. 2015)	Woven (552 g/m2)	Epoxy	51.28	2.74	24.	1.1
(R2)Woven kenaf (Treated NaOH)(Yahaya, Sapuan et al. 2015)	Woven (552 g/m2)	Epoxy	21.76	0.72	27	2.9
(R3)Woven kenaf (Yahaya, Sapuan et al. 2014)	Woven (552 g/m2)	Epoxy	6.64	3.27	16.46	0.5
Kenaf/glass hybrid(Davoodi, Sapuan et al. 2010)	Unidirectional (sheet molding compound (SMC) process)	Epoxy	200-240	12	71.68	3.08
Kenaf/polyester (Roslan, Ismail et al. 2014)	Short fiber (mould compress process)	Polyester	-	-	26.5	1.3
Treated Kenaf (Reza, Jamaludin et al. 2014)	Unidirectional 10 vol.%	Epoxy	-	-	58	6.8
Treated Kenaf (Reza, Jamaludin et al. 2014)	Unidirectional 30 vol.%	Epoxy	-	-	12.4	14.4
Treated Kenaf (Reza, Jamaludin et al. 2014)	Unidirectional 40 vol.%	Epoxy	-	-	16.4	18.15
Warp woven kenaf(Azrin Hani, Chan et al. 2013)	Woven (617 tex)	Epoxy	47.8	-	-	-
Weft woven kenaf(Azrin Hani, Chan et al. 2013)	Woven (617 tex)	Epoxy	11.1	-	-	-
Kenaf/polypropylene (Wambua, Ivens et al. 2003)	Random mat 40 vol.%	Polypropylene	27	2	-	-
Kenaf/corn starch (Shibata, Cao et al. 2008)	Randomly distributed short (10 mm) 60 vol.% fibre	Corn starch	-	4.8	-	-
Kenaf/polyester (Aziz and Ansell 2004)	Unidirectional 64 vol.%	Polyester	123	13	-	-
Untreated Kenaf (Yousif, Shalwan et al. 2012)	Unidirectional	Epoxy	235.13	5.57	-	-
Treated Kenaf (Yousif, Shalwan et al. 2012)	Unidirectional	Epoxy	301.64	6.74	-	-

Previously, most of the studies done on the woven form of textile composites were on synthetic, rather than natural fiber (Azrin Hani, Chan et al. 2013). Moreover, the studies done on natural fiber usually employed random orientation and compressed mat (Azrin Hani, Chan et al. 2013). It is also noted that lesser attention has been devoted to evaluating the mechanical properties of woven kenaf thermoset polymer at various woven and stacked layer orientation. The effect of woven and stacked layer orientation on tensile and flexural properties of kenaf thermoset is reported in this chapter.

Besides that, the application of Structural Health Monitoring (SHM) system in turbine blade is becoming very important. In order to employ SHM system, the dynamics characterization of mechanical system on turbine blade should be determined as well. Modal analysis is a common way of testing conducted to study the natural frequency, mode shapes and damping percentage of a structure (Larsen, Hansen et al. 2002). One of the SHM sensor devices is Macro Fiber Composite (MFC). It can respond as a sensor and micro energy harvester as well. The flexibility features make it very suitable to bond on big and vibrating structures and also it exibites high electromechanical coupling coefficient (Hyun Jeong, Choi et al.). Besides, it solves several issues highlighted in monolithic piezoelectric transducers such as mechanical stability of the piezoelectric transducer under huge stress, brittleness (Park, Lee et al. 2008), electrical breakdown of the material under high fields and reduction in efficiency due to dielectric losses and depolarization (Daue, Kunzmann et al.). MFC type d_{31} is proposed as it produces a large strain and more energy for small applied forces (Ali and Ibrahim 2012). Moreover, MFC can produce electrical power up to 65% of input mechanical energy, better performance due to its thin layer and classified as bimorph structure which generates double output as compare to unimorph structure (Ali and Ibrahim 2012).

The micro energy harvester reacts from the vibrations occurring in the structure body. This concept is applied in several researches and innovations technology (Sodano, Inman et al. 2005; Anton and Sodano 2007; Erturk and Inman 2009) and shown promising results. MFC is very flexible and the properties have improved in terms of mechanical stability and brittleness (Sodano, Inman et al. 2005).

The factors influencing the amount of harvested energy are piezoelectric material, proof mass, gap of interdigitated electrodes, overlapping effect of resonance frequencies, and operating mode of piezoelectric conversion (Ralib, Nurashikin et al. 2009; M.H. Kahrobaiyan, M. Asghari et al. 2014). Theoretically, this effect can be further described via the equivalent linear spring mass system (Figure 1) and equation 1, 2 and 3.

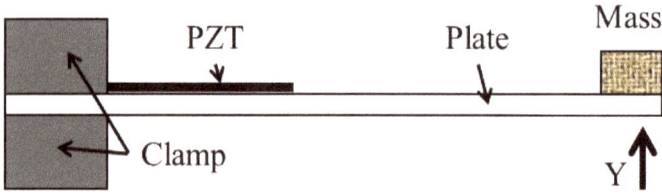

Figure 1. Cantilever beam with tip mass (Ralib, Nurashikin et al. 2009)

$$M\ddot{z} + C\dot{z} + Kz = -M\ddot{y} \tag{1}$$

$$\omega_n = \sqrt{\frac{K}{M}} \tag{2}$$

where $z = x\text{-}y$ is the net displacement of mass, M is the lumped mass, K is the spring constant, C is the damping coefficient and ω_n is natural frequency (Ralib, Nurashikin et al. 2009). The power generated from piezoelectric is correlated to natural frequency or resonance. The power output model at resonance can be written as

$$P_{max} = \frac{m\gamma^2\omega^3}{4\zeta} \tag{3}$$

where m is the seismic mass, γ is the amplitude vibration, ω is the system resonance frequency and ζ is the relative damping ratio (Ralib, Nurashikin et al. 2009). Therefore, the natural frequency of the beam or part should be identified. Hence, maximum power output can be gained and the range of natural frequency of the structure.

Most of the findings show that the MFC is bonded on the surface of the structure only (Hyun Jeong, Choi et al.; Daqaq, Stabler et al. 2009; Song, Choi et al. 2009). Hence, the combination of MFC in turbine blades could give new insights especially the effect of bonded MFC on the turbine blades and will be discussed in the next chapter. This chapter is reporting the bonded MFC effect technique on woven and unwoven kenaf plates via modal impact testing for micro energy harvested of VAWT application.

This chapter presents the best woven layer and stacked layer orientation for kenaf plate and turbine blade application. A preliminary report on MFC bonding technique in kenaf plate is conducted as well.

2.   Selection of woven and stacked layer orientation

Kenaf short fiber and fiber Kenaf yarns were utilized for this research. Kenaf yarns and kenaf short fiber was supplied by Juteko Co. Ltd., Bangladesh.The yarn fineness value is recorded as 300 Tex. Tex is a textile unit used to define and measure yarn and is equal to grams per kilometer. The matrix used was EpoxAmite 100 (Smooth-on) and cured with 102 Medium Hardener (Smooth-on) hardener. The physical properties of EpoxAmite 100 and 102 hardener combination from the manufacturer are shown in Table 2.

Table 2 Physical properties of combination of EpoxAmite 100 and 102 hardener

Physical properties	PSI	Pa
Flexural Strength (ASTM D790)	12,220	84.25×10^6
Flexural modulus (ASTM D790)	423,000	2.91×10^9
Ultimate tensile strength (ASTM D638)	8,180	56.40×10^6
Tensile modulus (ASTM D638)	450,000	3.10×10^9

Epoxy resin was employed as it is light weight and causes less damage to the manufacturing equipments and has better mechanical properties compared to other resin (Yousif, Shalwan et al. 2012). Kenaf yarn was weaved manually using a self designed handloom. Figure 2 shows the self-designed handloom employed in weaving inbalanced plain woven kenaf. Weft and warp setup for woven kenaf was 11 ends per centimeter or 28epi (ends per inch) and 3 picks per centimeter or 8ppi (picks per inch) respectively as shown in Figure and Figure .

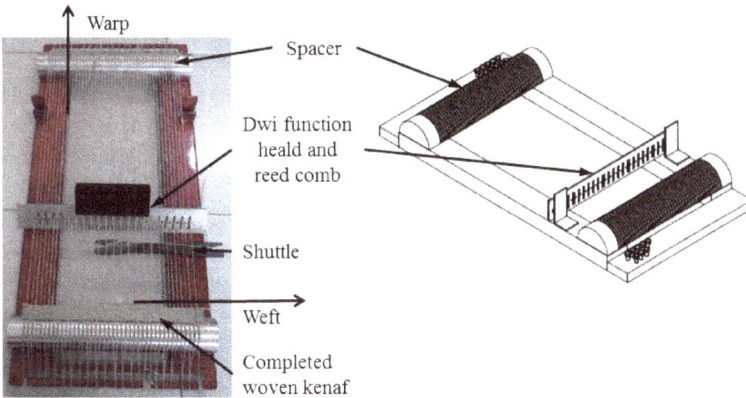

Figure 2 Self -design handloom

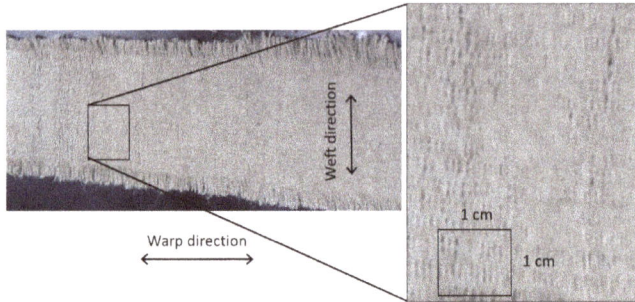

Figure 3 Completed woven kenaf

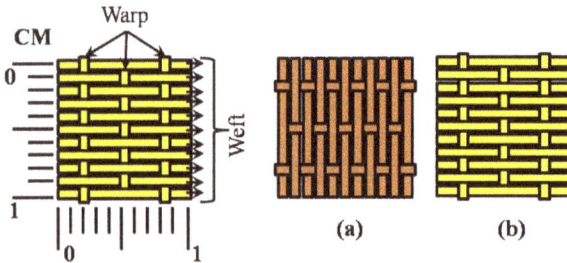

Figure 4 Schematic of kenaf woven orientation (a) Orientation A, (b) Orientation B

The kenaf weaving process can produce a maximum size of 20 cm x 25 cm of woven kenaf. The fabrication of woven and unwoven kenaf plates were conducted via a vacuum infusion process. The ratio between kenaf and epoxy resin is 30:70 following the ratio used by Yousif et al. (Yousif, Shalwan et al. 2012) and Yahaya et al. (Yahaya, Sapuan et al. 2014).The resin was prepared in a combination of epoxy to hardener of 100g:28.4g as recommended by the manufacturer (**Error! Reference source not found.**). Three layer of woven kenaf were stacked at average 70 g in overall weight. For unwoven kenaf, 70 g kenaf short fiber is setup on the fabrication mould. In this testing, there were five different samples fabricated. Four woven kenaf samples were differentiated from woven waft and weft orientation as shown in Table 3 and Figure 5 for sample 1, Figure 6 for sample 2, Figure 7 for sample 3 and Figure 8 for sample 4. Unwoven kenaf is fabricated as a control sample. The woven and stacked layer orientation is presented in Table 3.

Table 3 List of samples employed at different orientation for mechanical properties testing

Sample	Orientation
S1	A / A /A
S2	B / A / B
S3	A/ B /A
S4	B/ B / B
S5	Unwoven kenaf

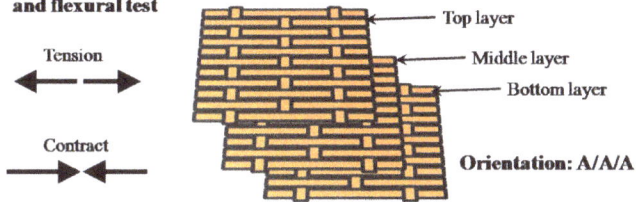

Figure 5 Sample 1: Orientation A/A/A

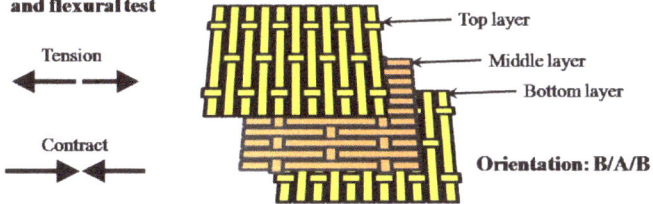

Figure 6 Sample 2: Orientation B/A/B

Force direction for tensile and flexural test

Tension

Contract

Top layer

Middle layer

Bottom layer

Orientation: A/B/A

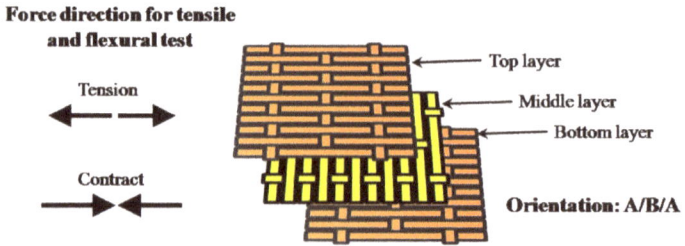

Figure 7 Sample 3: Orientation A/B/A

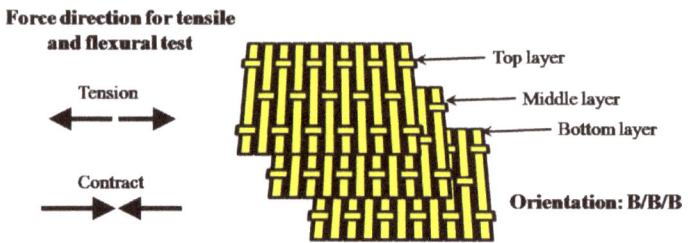

Force direction for tensile and flexural test

Tension

Contract

Top layer

Middle layer

Bottom layer

Orientation: B/B/B

Figure 8 Sample 4: Orientation B/B/B

3. Mechanical testing and scanning image microscopy analysis

The flexural test is conducted from a 3-point loading using a Instron 3365 testing machine according to ASTM D 790-03. The rectangular samples of dimension 127mm x 12.7mm were cut using a circular saw. The tests were conducted at the cross head displacement rate of 45 mm/min. The reports were automatically generated by the software embedded in the machine's computer. For each sample, five specimens were tested at room temperature and the average was taken as a final result.

Tensile test was conducted to determine the stress–strain behaviour of the woven kenaf composite. The test was carried out using a 810 Material Testing Machine (MTS) based on ASTM D 3039 on plates with a size of 200mm x 25mm and 3.5-4.5mm sample thickness for each composite. The samples were carefully cut from the laminate using a wheel saw and finished to the accurate size. 500 psi grip pressure was applied to the samples. A standard head displacement at a speed of 2 mm/min was applied. For each sample, five specimens were tested and average results were recorded.

The results were compared with previous research findings to identify the effect of fiber orientation to the flexural and tensile properties of woven kenaf. In order to study the morphological feature of fibrematrix interface on woven kenaf and mechanical testing failure surface, the surfaces of the samples were examined using a scanning electron microscopy (SEM) (Hitachi model SU1510). Prior to the test, the samples were cut in to 10mm x 10mm and adhered on an aluminium plate. The image were focussed at 140 time magnification.

4.   Result and discussion

The flexural properties results are shown in Figure 9 to Figure 11. A flexural test is conducted to determine the strength of materials to withstand bending forces before the breaking point. The maximum load for each sample's orientation is depicted in Figure 9. It exhibit that S1 is experienced the highest maximum load. It is followed by S3, S2, S4 and S5. The density of weft yarn is believed to increase the strength of each layer. In woven kenaf, a low density in warp direction may contribute to the lower strength of kenaf composite as demonstrated in S4. Unwoven kenaf shows the lowest strength as short fiber kenaf has the lowest bonding among the fiber. Figure 10 exhibits the stress-strain curve for each sample. S1 exhibits the highest flexural strength, highest elongation break and highest flexural modulus.

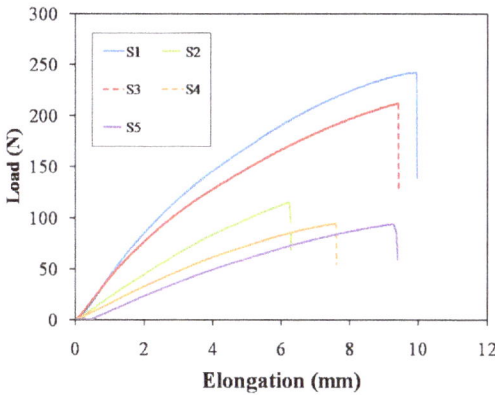

Figure 9 Flexural-Load elongation curves for kenaf composite

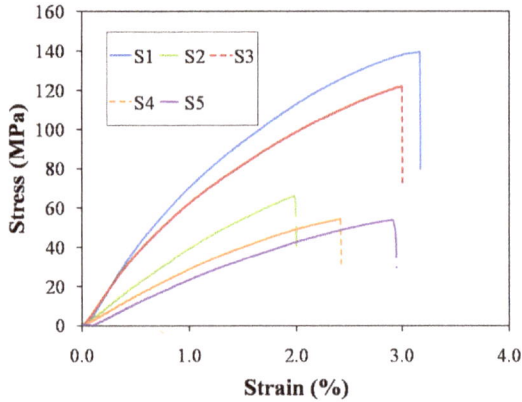

Figure 10 Flexural-Stress-strain curves for kenaf composite

The result for flexural strength and flexural modulus are shown in Figure 11. Generally, woven kenaf indicates a higher flexural strength and flexural modulus as compared to unwoven kenaf. S1 presents the highest flexural strength at 127.47 MPa. It is followed by S3, S2, S4 and S5 with 116.30Mpa, 60.98MPa, 48.07MPa and 42.67MPa respectively. There is a 9% increment on flexural strength between S1 and S3. Meanwhile, S1 woven kenaf shows 199% increment on flexural strength compared to unwoven kenaf. The flexural modulus for S1, S2, S3, S4, S5 is 7.5GPa, 3.7GPa, 6.6GPa, 3.1GPa and 2.7GPa respectively. S1 shows the highest flexural modulus compared to the lowest value, S5. Woven kenaf improved up to 178% as compared to unwoven kenaf. The difference in increment between S4 and S2 is 19%. Significant increase in percentage occurred between S2 to S3 and S3 to S1 which are 78% and 14% respectively. In flexural test, a compressive mode occurred on the top layer while the bottom layer experienced tensile force (Hani, Ahmad et al. 2013).

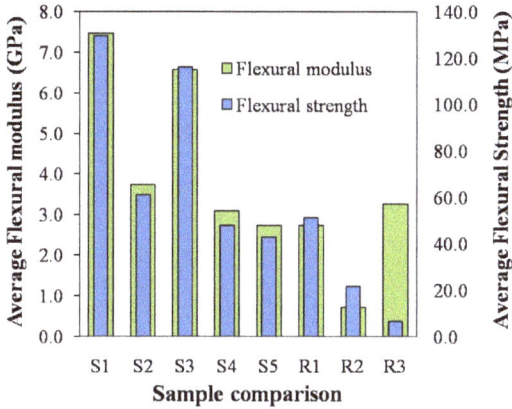

Figure 11 Flexural properties of composite

In the flexural test, the compressive mode was found to be the main cause of failure of the top layer sample (Park and J. Jang 1997; Hani, Ahmad et al. 2013). The difference in fiber orientation in each layer may cause the diffrent flexural strength of each sample. The result indicates that woven kenaf has shown significant improvement compare to the unwoven kenaf. Besides, the fiber orientations clearly effect the material strength. The A orientation exhibit a higher strength compared to the B orientation. This may be due to a higher yarn density which can support and resist the force exerted at transverse direction of the sample. S1 indicates the highest strength as the sample is supported by three layers of the A type. This is followed by S3 which covers two layers of A type orientation. S2 is higher than S4 as it has one A type layer compared to S4 which only employs B type layers.

Furthermore, flexural strength and flexural modulus comparison with resin properties supplied by the manufacturer as shown in Table 2 reveals positive findings. Flexural strength and flexural modulus shows increment percentage to 51% and 157% respectively when compared to S1 as the highest value. It shows that fiber orientation and layer arrangement in S1 perform excellent improvement especially in flexural strength and modulus. S1 is preferred for fabrication selection in the next experiment.

Tensile properties are illustrated in Figure 12 to Figure 14. Figure 11 exhibits the maximum load resisted by the material before it breaks. S3 shows the highest elongation resistance and S1 presents the highest load resistance. Figure 13 shows the stress-strain curve which indicates the ultimate tensile strength at maximum graph and tensile modulus for graph gradient. Tensile modulus and strength represents the ability of material to resist with any tensile deformation. The results for tensile strength and tensile modulus are reported in Figure 14.

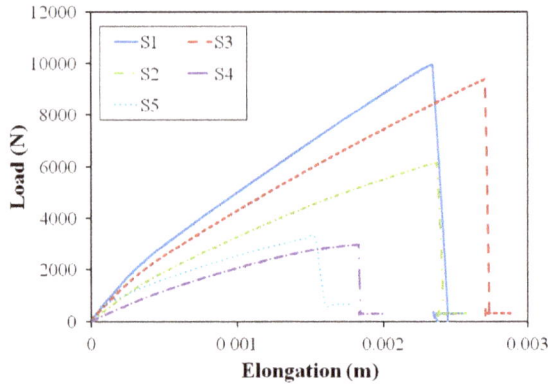

Figure 12 Tensile -Load elongation curves for kenaf composite

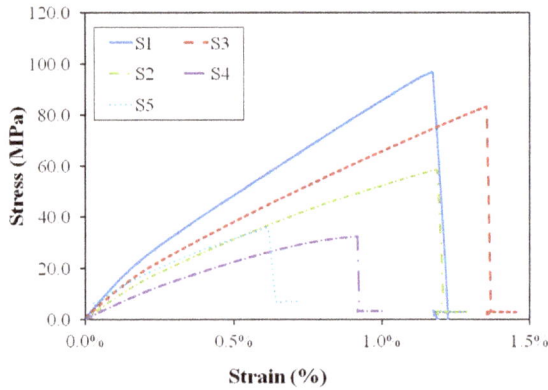

Figure 13 Tensile- Stress-strain curves for kenaf composite

Tensile modulus and tensile strength illustrate similar trends. S1 exhibits the highest tensile strength and tensile modulus, followed by S3, S2, S5 and S4. The values for tensile strength are 94.3MPa, 77.7MPa, 55.9MPa, 34.0MPa and 28.1 for S1, S3, S2, S5 and S4 respectively. Tensile modulus for S1, S3, S2, S5 and S4 are 8.1GPa, 6.2GPa, 4.9GPa, 4.4GPa and 3.5GPa respectively. Generally, woven kenaf is better than unwoven kenaf except for S4. An increase in thickness may cause decrease on tensile strength irrespective to fiber orientation (Yahaya, Sapuan et al. 2014). The average thickness of S4 is 3.7mm while S5 is 3.4. An 8% different may cause decrease of tensile strength for S4 as compared to S5. S1 shows an improvement of tensile strength and modulus compared to S5 at 177% and 84% respectively. In woven kenaf, fiber orientation clearly exhibit a huge influence on the tensile properties. S2 improved 99% for tensile strength and 40% for tensile modulus respectively compared to S4. Improvement from S2 to S3 about 39% and 26% for tensile strength and tensile modulus respectively. Lastly, S1 is improved compared to S3 at 21% and 31% for tensile strength and tensile modulus respectively. This shows that the highest number of yarn in each woven area exhibits better tensile and flexural properties.

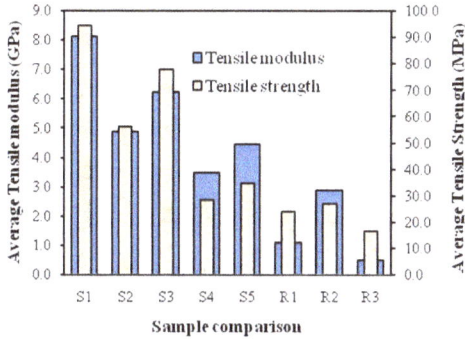

Figure 14 Tensile properties of composite

Besides that, the ultimate tensile strength and tensile modulus results are compared with resin properties shown in Table 2. Ultimate tensile strength and tensile modulus shows an increment up to 67% and 161% respectively when compare to S1. Similar to flexural test analysis, it supports the improvement produce due to fiber orientation and layer arrangement orientation.

Figure 15 and Figure 16 depict the fiber morphology of woven kenaf and unwoven kenaf respectively. The analyses were conducted on flexural samples which had undergone a compress effect on the top layer and a tension effect on the bottom layer. Generally, both figures show that interaction between fiber and matrix is a bonded interaction without resin penetration occurring into the fiber. There are obvious gaps seen in both figures indicating poor bonding interaction between fiber and matrix resin.

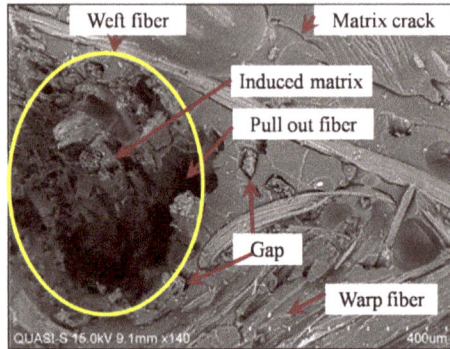

Figure 15 SEM analysis on woven kenaf

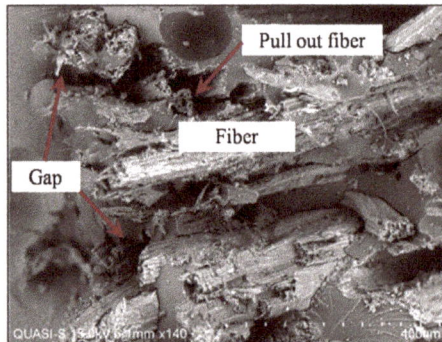

Figure 16 SEM analysis on unwoven kenaf

Figure 15 shows that the resin matrix is induced into the kenaf fibers gap. Hence, it gives additional strength and a bigger bonding surface area between resin and fiber as compare to unwoven kenaf. Figure 16 shows that the fiber's size are mixed and randomly arranged. The big fiber may decrease the bonding surface area hence decreasing the bonding strength. The fracture mechanisms occurring for woven and unwoven are pulled out, detached and debonded from epoxy matrix as depicted in the figures. The matrix crack occurred is revealed in the composite as shown in Figure 15.

Besides that, further analysis conducted on type A (Sample 1) and type B (Sample 4) orientation. Figure 17 shows the SEM analysis at three different points near the fiber. The microstructure at break point was investigated and elucidated. It shows that the resin is properly diffused inside the fiber. Hence make the strength increased. Furthermore, the fracture mechanism occurred are pulled out and detached. No gap or air bubble effect on the composite is noticed.

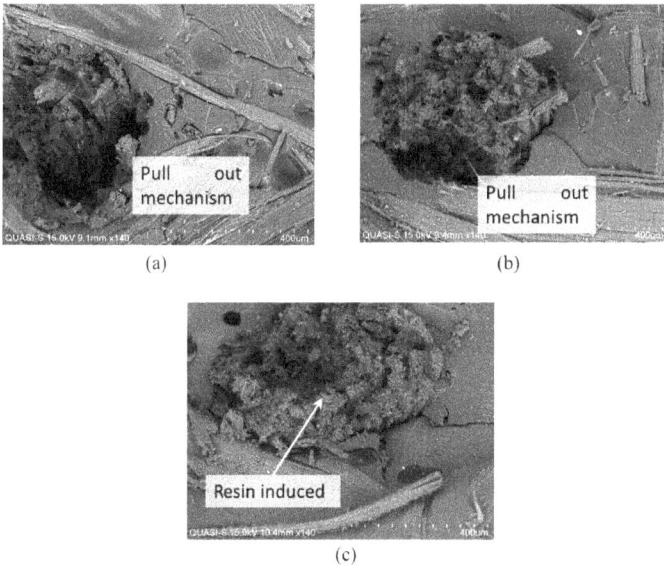

(a)   (b)

(c)

Figure 17 SEM analysis on sample 1 (Type A)

Figure 18 shows the SEM analysis on the type B or Sample 4. Generally, gap is observed in the picture. It may cause to the strength reduction. The fracture mechanism occurred exhibit pulled out and debonded mechanism. Air bubbles can be spotted at several areas in the resin region.

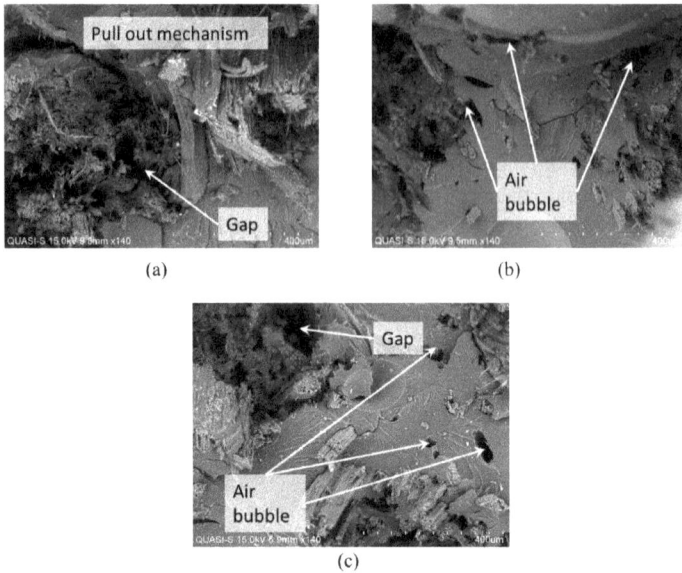

(a)

(b)

(c)

Figure 18 SEM analysis on sample 4 (Type B)

Several previous papers on woven kenaf thermoset polymer indicate findings on tensile and flexural properties as shown in Table 1. Three results from Table 1 namely R1, R2 and R3 are compared in Figure 11 and Figure 14. The main difference between group R and group S is on fiber orientation, yarn size and fabrication process. Generally, group S shows better result as compared to group R for tensile and flexural properties. The flexural strength for S1 has improved by 152% compared to R1 (the highest in group R). Meanwhile, there was an improvement of 127% for S1 compare to R3 (the highest in group R) for flexural modulus. The similar trend occurred in tensile properties. S1 improved to 250% and 179% for tensile strength and tensile modulus respective as compared to the highest value in group R. This result exhibits that small size of yarn produce a higher yarn density for each woven area given a better finding on tensile and flexural properties. Indirectly, vacuum infusion process is reported to give many more advantages compare to a hand lay-up process for woven kenaf thermoset polymer. However, unidirectional type fiber tends to exhibit outstanding mechanical properties compared to woven kenaf as shows in Table 1 (Aziz and Ansell 2004; Yousif, Shalwan et al. 2012 ; Reza, Jamaludin et al. 2014). The volume of fiber in each area of kenaf may effect and influence the mechanical properties of kenaf polymer (Reza, Jamaludin et al.

2014). Higher volume fiber exhibits better flexural and tensile properties for natural fibers up to the optimum level.

5.   Macro fiber composite (MFC) bonding technique effect on the woven kenaf plate

The orientation AAA or sample 1 was utilised for the woven kenaf plate fabrication. Three layers of woven kenaf with dimensions of 200mm x 250mm were stacked at an average weight of 70 g. For the unwoven kenaf, 70 g of kenaf short fibre was setup on the fabrication mould. For the embedded MFC, this was laid at the specific layer and dimension of kenaf as shown in Figure . Bonded MFC was attached onto the surface after curing, assuming that the bond is effectively adhered so that the stress is effectively transferred between the two surfaces of the kenaf plate.

After curing, the plate was cut into specific dimension as shown in Figure 20 and Figure 21. Four plates were fabricated following the sample type stated in Figure 21 and Table 4. A reference plate was fabricated from woven kenaf as well via the vacuum infusion technique. The same plates were employed for the vibration test after the modal testing was conducted.

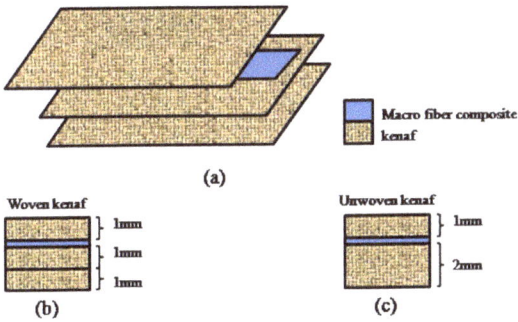

Figure 19 Schematic of MFC location and thickness of the kenaf layer, (a) Schematic of the kenaf layer and MFC for woven kenaf plate, (b) thickness of the kenaf layer for woven kenaf plate, (c) thickness of the kenaf layer for unwoven kenaf plate.

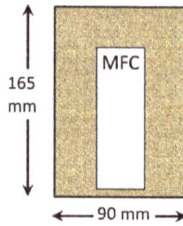

Figure 20 Dimensions of Kenaf plates for modal testing

Figure 21 Sample for modal testing (a) MFC bonded onto unwoven kenaf, (b) embedded MFC in unwoven kenaf, (c) MFC bonded onto woven kenaf, (d) MFC embedded in woven kenaf

6. Experimental method

6.1 Modal testing

The electricity induced by the MFC was influenced by the deformation in the plate structure. Theoretically, the high peak voltage output induced during the resonance of the structure is due to maximum deformation (Tien and Goo 2010). Modal testing analysis is employed to assess the natural frequency or resonance of any structure. In this experiment, a simple modal testing approach was utilised known as single reference testing. A hammer impact technique is a common application for single reference testing and has been selected due to this being the easiest method. The impact hammer modal analysis excitation method was utilised to analyse the natural frequency of each plate and the effect of the MFC bonded technique on the plate structure. Pabut et al.'s (Pabut, Allikas et al. 2012) modal testing methodology is referred to in this experiment. The sample characteristics and experimental setup are shown in Figure 22 and Table 5 respectively.

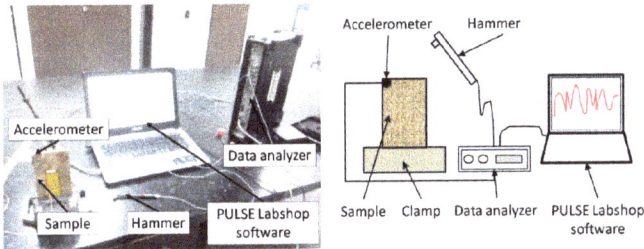

Figure 22 Experimental setup for impact hammer excitation in Modal Testing

Table 4 Sample characteristics for modal testing analysis

Sample	Parameter	
	Bonding type of MFC patch	Woven type of kenaf
1- Embedded woven plate	Embedded inside the Kenaf Plate	Woven kenaf
2- Embedded unwoven	Embedded inside the Kenaf Plate	Unwoven kenaf
3- Bonded woven plate	Bonded on the Kenaf Plate surface	Woven kenaf
4- Bonded unwoven plate	Bonded on the Kenaf Plate surface	Unwoven kenaf
5- Woven plate	-	Woven kenaf

The samples were rigidly positioned and clamped. Nine static points were identified based on grid point selection for hammer excitation as shown in Figure 23. The grid was setup in equal distance to obtain clear mode shape illustration. Figure 23 shows that the accelerometer (B&K4517) was mounted on the corner of the plate assuming the maximum deformed locations on the plate.

The hardware devices required in this modal testing are: Fast Fourier Transform (FFT) analyser (Model Brüel & Kjær, Denmark- B&K 3560 C) to compute the Frequency Response Functions (FRF). Laptop (for PULSE Labshop software and ME'ScopeVES software) to identifying the modal parameter and mode shapes animation. Accelerometer (B&K4517) to capture and measure the response acceleration at a fixed point and direction and it also operates as an output reference. A B&K 8204 modal impact hammer was used as an exciter input force on the structure which is equipped with a load cell to record the load induced in the structure.

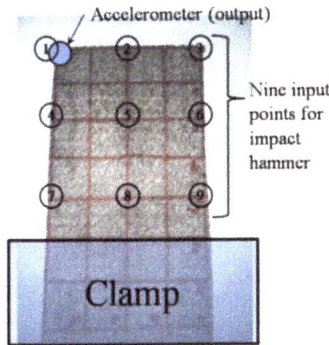

Figure 23 Location of accelerometer, clamped area and nine points for hammer excitation

An impact hammer was exerted on each point five times to obtain satisfactory, accurate data and to minimise errors. An FFT analyser was equipped with a spectrum averaging capability to accept or reject the hit from the hammer due to human error. The error indicator is denoted by coherence graph information. The best results are recorded if the coherence line shows a smooth line without distortion. The response data was acquired using the Brüel & Kjær PULSE multi analyser platform (only the 2-Channel FFT analyser was used in this impact hammer analysis) which also provides FFT based validation tools and computes Frequency Response Functions (FRF).

The PULSE Labshop software takes the raw time data and, based upon this, estimates the natural frequencies, the mode shapes and the damping percentage via ME'ScopeVES post-processing software. The software is setup to capture the frequency range from 0 to 800 Hz. Further analysis was only focused on an audible frequency range, typically 30-400 Hz due to potential noise problems occurring in the wind turbine blades (Jenq, Hwang et al. 1993).

The results displayed in the FRF as a sample are shown in Figure 24, and were analyzed via the Curve fitting method in PULSE Labshop software and ME'ScopeVES post-processing software. Curve fitting is a matching process which occurs between the mathematical expression and the empirical data points. This is conducted by minimising the squared error between the analytical function and the experimental data. FRF depicts the input-output relationship between two points as a function of frequency. In this analysis, the log magnitude was recorded from the accelerometer data divided by the force exerted by the hammer output (m/s^2)/ input (N) as a function of frequency. Each peak from the graph represents the mode or natural frequency of the structure. The deformation shape was different at various natural frequencies and this is evaluated in mode shape analysis. The report outline from modal testing is very useful for analysing the correlation of the MFC bonding technique in the kenaf plate and the natural frequency of the plate.

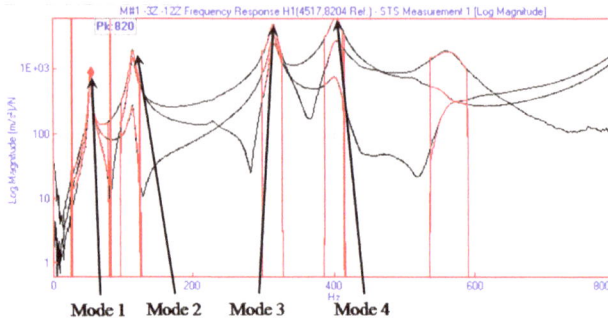

Figure 24 Example of Frequency Response Functions (FRF) and natural frequency identification

6.2   A plate vibration test

A vibration test was conducted to analyse the performance of the MFC energy harvester at the respective range of frequencies for wind turbine application. The prepared plates were clamped to the jig at one end and were left free at the other hand as shown in Figure 12. The tip of the shaker was applied to the free end of the plate to simulate the vibration at the respective range of frequencies. The details of the equipment employed in this experiment are summarised in Table 5.

Table 5 Equipment for vibration testing

Shaker	Amplifier	Vibration controller	Oscilloscope
Modal Shop : model: 2007E	Modal shop: 2100E21-100	Hardware: LMS: SCM 202	Picoscope 3206B
To vibrate the plate	To amplify the signal from the controller to shake at 3W power	Software: LMS To control the frequency of the shaker	To record the voltage induced from MFC

The vibration test setup follows the concept developed by Sodano and Ali (Sodano, Inman et al. 2005; Ali and Ibrahim 2012; Wang, Yang et al. 2012). The amplifier was set at 3 watt. The range of frequencies in this experiment was selected to be 20 Hz to 100 Hz referring to mode one of the natural frequency and 120 Hz to 200 Hz for mode two of the natural frequency of the plates. This range was proposed to ensure that it occurred at least at 5 Hz before and after the natural frequency following the Tien and Goo technique (Tien and Goo 2010). The natural frequency values were obtained from the modal testing experiment. The voltage generated, as induced from the MFC, was transmitted to a smart power harvesting module (EH-CL 50 from smart-material) before connection to the oscilloscope for voltage recording. EH-CL50 was specially developed for P2 type MFC in low frequency/intermittent harvesting applications. Hence, no additional harvesting circuit was needed to harvest the induced voltage. The operation is based on capacitive energy extraction.

The experimental setup between the plate and the shaker is shown in Figure 25. The shaker frequency and amplitude were controlled by a frequency controller and an amplifier respectively. The shaker frequency was set at 20, 40 and 60 Hz for first mode analysis, and 80, 100,120 and 140 Hz for second mode analysis. Four different plates were tested. The experimental parameters are summarised in Table 6. The voltage RMS generated was recorded via an oscilloscope (PICOSCOPE) and saved on a computer.

Table 6 Summary of experimental parameters for vibration testing analysis and micro energy harvesting testing on the prototype of the wind turbine

Factor	Experimental Condition Levels	
	1	2
A-Bonding type of MFC patch	Embedded inside the Kenaf Plate	Bonded on the Kenaf Plate surface
B-Woven type of kenaf	Woven	Unwoven
C-Input frequency (Hz)	20, 40 and 60(first mode) 80, 100, 120, 140(second mode)	

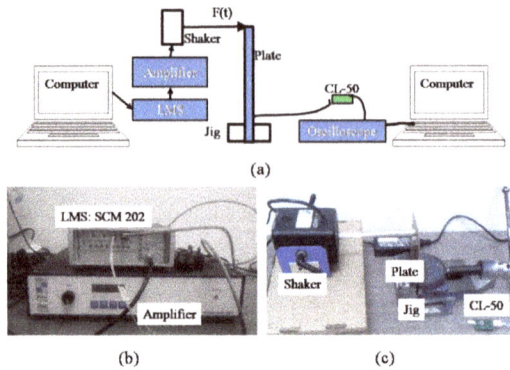

Figure 25 Experimental setup for the vibration test, (a) Schematic drawing, (b) amplifier and LMS shaker controller and (c) shaker and clamped plate.

6.3   Statistical analysis via Taguchi method

Phadke (1989) and Taguchi (1996) explained that a robust design is an engineering methodology for optimising the product and process conditions which are minimally sensitive to the various causes of variation and which produce a high quality product with low development and manufacturing costs. The Taguchi method is one of the applications employed so that the significance of the factor, or multiple factors, that affect the particular process's performance can be statistically determined.

The Taguchi methods are statistical methods initially developed by Genichi Taguchi to improve the quality of manufactured goods. More recently, the techniques have been used in scientific and engineering experiments since they allow for the analysis of many different parameters without a prohibitively high number of experiments.   Many researchers now apply robust design as a tool to achieve quality engineering in many fields.

Taguchi proposed that engineering optimisation should be looked at in terms of a three-step approach: system design, parameter design, and tolerance design (Phadke 1989; Taguchi 1990). In the system design, the aim is to produce a basic, functional prototype design which is inclusive of the product design stage and the process design stage. Several factors, such as material selection, component selection, and tentative product parameter values, are involved in the product design stage; whereas the process of the design stage involves, for example, the analysis of processing sequences, and the selection of production equipment and tentative process parameter values. The optimum quality and cost are not related to this system design.

On the other hand, the parameter design approach involves the step to optimise the process parameter values for improving the quality characteristics as well as to ascertain the product parameter values under optimum conditions. Besides this, the output from the parameter design is considered to be less sensitive to the environmental condition variation and other noise factors.

Lastly, the objective of tolerance design is to determine and investigate the tolerance around the optimal settings recommended by the parameter design. Tolerance design is required if the reduced variation obtained by the parameter design does not meet the required performance, and it involves tightening the tolerances on the product parameters or process parameters for which variations result in a large negative influence on the required product performance. Parameter design is the approach that emphasises optimising high quality products, as well as reducing the product cost. In order to achieve the objectives of this study, parameter design using the Taguchi method is employed. Furthermore, Taguchi offers an experimental design wholly based on statistical design as a tool which is less sensitive to noise factors. The two major tools used are: the Signal to Noise (S/N) ratio, which measures quality with the emphasis on variation, and orthogonal arrays, which accommodate many design factors simultaneously. The significance of the factor or multiple factors that affect the machining quality performance could be determined in a very short time when this technique is employed.

Orthogonal arrays were adopted to determine the appropriate design of the experiments which can be carried out with the lowest number of experiments. This technique has a balanced property in which each factor setting occurs the same number of times for every setting of all the other factors in the experiment. Orthogonal arrays allow researchers or designers to study many design parameters simultaneously and can be used to estimate the effects of each factor independent of the other factors. Therefore, information about the design parameters can be obtained with the minimum of time and resources (Antony and Kaye 1999).

The method for calculating the S/N ratio response is designed in three different modes depending on whether the quality characteristics are: smaller the better, larger the better, or nominal the better(Taguchi 1990).The equations for calculating the S/N ratio are listed below:

For the smaller the better characteristic (in dB):

$$S\!/\!_{N} = -10 \log \frac{1}{n} \left(\sum y_i^2 \right) \tag{4}$$

For the larger the better characteristic (in dB):

$$S_N = -10 \log \frac{1}{n} \left(\sum \frac{1}{y_i^2} \right) \tag{5}$$

For the nominal the better characteristic (in dB):

$$S_N = -10 \log \frac{\bar{y}}{s_{y_i}^2} \tag{6}$$

Where $\bar{y}$ is the average of the observed data, $s_{y_i}^2$ is the variance of y, n is the number of observations and y_i is the observed data. The S/N ratio values function as a performance measurement to develop processes insensitive to noise factors.

The degree of predictable performance of a product or process in the presence of noise factors could be defined from the S/N ratio values. For each type of characteristics, with the above S/N ratio, the higher the S/N ratio, the better the result. The S/N ratio is presented in a response graph and table. The analysis of the S/N ratio value was conducted using Minitab software (version 16.0). The experimental designs were determined in the software. This depends on the number of factors and levels.

The Taguchi design of plate vibration experiment was divided into two. For the first mode, the analysis design was L$_{36}$ orthogonal array. The two factors with two levels and one factor with three levels as shown in Table 8. Meanwhile, for the second mode was L$_{18}$ with three factors. The two factors with two levels and one factor with four levels as shown in Table 10. Voltage RMS was considered as output result. Appendix 2 shows the available design in Taguchi Design of experiment for the research conducted.

7. Result and discussion

7.1 Modal testing analysis

Natural frequencies for different bonded and woven MFC types are shown in Table 7. The plates may optimally deform and deflect during resonance. Hence this can influence the performance of the energy harvesting generated from the MFC. The analysis of the natural frequency in kenaf plates is focused on the range 0 Hz to 400 Hz as indicated in Chapter three, following the methodology of Jenq et al. (Jenq, Hwang et al. 1993) for wind turbine blades. The reference plate, embedded unwoven plate and embedded woven plate have recorded a single natural frequency at the suggested bandwith ranges. Mode four for the embedded woven plate is 404 Hz and for the embedded unwoven plate it is

615 Hz. While bonded unwoven and bonded woven plates generate two resonance frequencies in the range 0 Hz to 400 Hz. Generally, the bonded MFC patch reduces the natural frequency of the plate as compared to the reference plate for each mode.

Table 7 Natural frequencies of various kenaf plates for the first four modes

Mode	embedded woven plate (Hz)	embedded unwoven (Hz)	bonded unwoven (Hz)	bonded woven plate (Hz)	Reference (Hz)
1	54.3	42.5	40.3	29	56.8
2	114	139	125	92	127
3	316	266	225	168	369
4	404	651	388	310	473

Theoretically, the main factors influencing the natural frequency of a structure are determined by the mass, the stiffness damping properties and the boundary condition of the structure. Any changes in these factors may change the natural frequency of the structure and the mode shapes as well.

An analysis of the mode shapes for embedded woven plate, embedded unwoven plate, bonded unwoven plate and bonded woven plate is shown in Figure 26, Figure 27, Figure 28 and Figure 29 respectively. In the embedded woven plate (Figure 26), the first mode is deflected by the bending effect, the second mode is deflected by torsion, the third mode is deflected by bending at a different axis and the fourth mode is a combination of both bending and torsion.

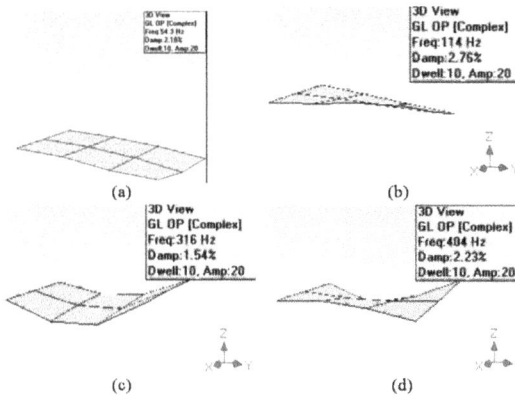

Figure 26 Mode shapes for the embedded woven plate (a) first mode, (b) second mode, (c) third mode, (4) fourth mode

The deflection shapes for embedded unwoven plate (Figure 27) are defined as bending, torsion, bending at a different axis and a combination of torsion and bending for the first, second, third and fourth modes respectively.

In addition, the mode shapes for the bonded unwoven plate are shown in Figure 29. The first mode is deflected by bending, the second mode is by torsion, the third mode is by bending and the fourth mode is a combination of torsion and bending. For the bonded woven plate (Figure 29), the deformation shape for each mode occurred at the same condition as the bonded unwoven plate.

This analysis shows the type of deformation that occurred in the kenaf flat plates. The deformation is influenced by the weight and the plate stiffness. The types of deformations are different at different ranges of vibration frequency. This information may guide researchers to understand the structural deformation pattern at specific ranges of vibration. It also gives ideas about suitable locations for the sensor. Hence, the SHM system and the energy harvesting system can be properly employed and utilised in their best conditions.

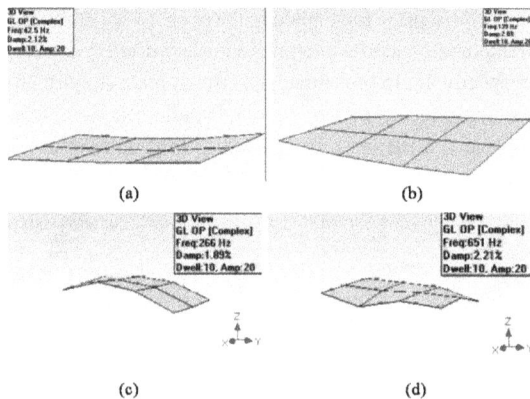

Figure 27 Mode shapes for embedded unwoven plate (a) first mode, (b) second mode, (c) third mode, (4) fourth mode

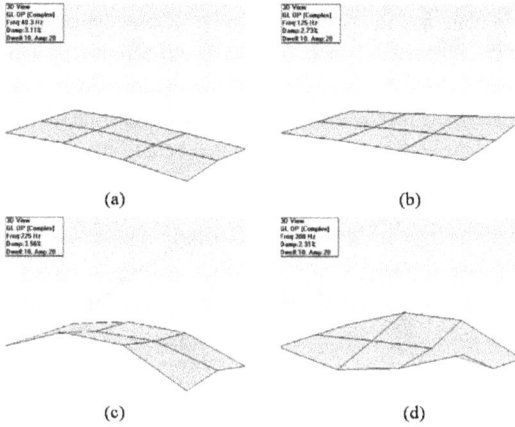

Figure 28 Mode shapes for bonded unwoven plate (a) first mode, (b) second mode, (c) third mode, (4) fourth mode

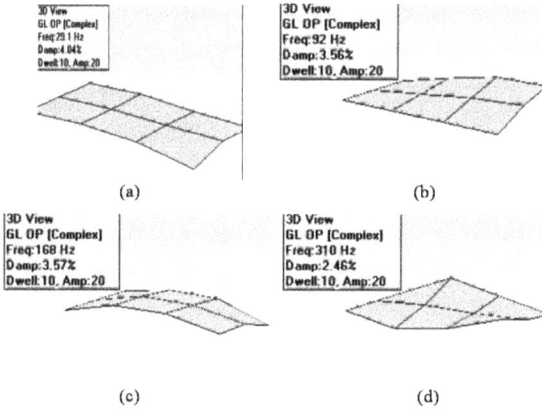

Figure 29 Mode shapes for bonded woven plate (a) first mode, (b) second mode, (c) third mode, (4) fourth mode

An analysis of the damping percentage has also been conducted. Figure 30 shows the damping percentage for each of the kenaf plates. The values are considered to be small and do not exceed 4%. The trend and damping values are almost the same for the embedded woven plates and the embedded unwoven plates. Besides this, the trends for the bonded unwoven plate and the bonded woven plate are almost the same. However, at mode one and two, the difference is about 1%. The deflection type and the bonded MFC type may contribute to the value and trend of the damping percentage. The reference plate shows the lowest value for the damping percentage, and the trend is the same for the embedded MFC plates for both woven and unwoven kenaf fibre.

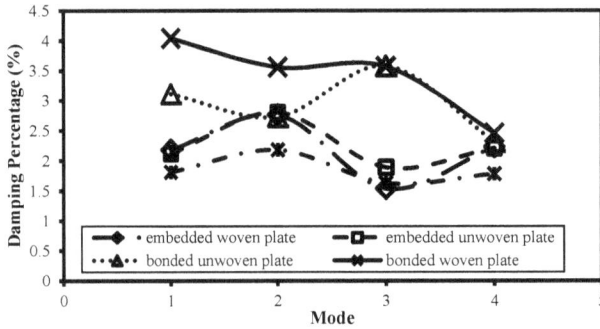

Figure 30 Damping percentage for kenaf plates at each mode

Modal testing shows three reports which are: the natural frequency of the kenaf plate, the mode shapes and the damping percentage. The results indicate that for particular kenaf woven and unwoven plates, bonded MFC influenced the damping percentage significantly. The deformation types show a similar condition for the first mode to the fourth mode. The deformations are: bending for the first mode, the second mode is torsion, the third mode is by bending and the fourth mode is a combination of torsion and bending.

7.2   Plate vibration test

For first mode analysis, the result is presented in Figure 31. Figure 32 displays the average value of the voltage RMS in each frequency. All results exhibit a voltage range between 100mV to 600mV. Embedded woven plates show the lowest results, which are 103.6mV, 120.3mV and 110.9mV for 20Hz, 40Hz and 60Hz respectively. For the embedded unwoven plates, the results are 546.9mV, 157.7mV and 149mV for 20Hz, 40Hz and 60Hz respectively. The bonded woven results are 448.1mV, 472.7mV and 436.7mV for 20Hz, 40Hz and 60Hz respectively. Meanwhile for bonded unwoven plates, the results are 392.2mV, 119.9mV and 115.3mV for 20Hz, 40Hz and 60Hz respectively.

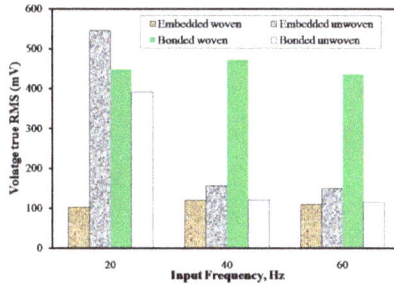

Figure 31 Voltage RMS result from vibration plates in first mode analysis

S/N ratio vibration analysis for the first mode is reported in Figure 32 and Table 9. Table 8 shows the experimental parameters as a reference. Further analysis was conducted using the response table of the S/N ratio and the response graph of the S/N ratio to identify the most influential factor and the optimum level in each factor to obtain a higher voltage response.

Table 8 The experimental parameters for first mode plate vibration testing analysis

Factor	Experimental Condition Levels		
	1	2	3
A-Bonding type of MFC patch	Embedded inside the Kenaf Plate	Bonded on the Kenaf Plate surface	0
B-Woven type of kenaf	Woven	Unwoven	0
C-Input frequency, Hz	20	40	60

Table 9 shows that the bonded type is proposed as the most influential factor, followed by fibre type and frequency. This result refers to the Delta item. Delta is the difference between the minimum and maximum S/N ratio for each of the factors. The higher value means that it is considered as the most influential factor above the others. Rank refers to the order of the influencing factor in each analysis. Number one is considered as the most influential factor followed by number two and so on.

The optimum level is identified from the highest S/N ratio value in each factor. For the bonding type factor, level 2 (which is bonded MFC) is proposed as the optimum level. While for fibre type, level 1 (which is the woven type) is proposed as the optimum level.

Lastly, the optimum frequency is suggested at level 1 which is 20 Hz. Figure 32 depicts that the optimum parameters necessary to achieve a higher voltage RMS is a combination of bonded MFC and woven kenaf fibre and this operates close to natural frequency.

Table 9 Response Table for S/N Ratio in first mode analysis

Level	Factors		
	Bonding type	Fibre type	Frequency
1	44.17	48.53	47.75
2	48.6	44.24	46.18
3			45.23
Delta	4.42	4.3	2.52
Rank	1	2	3

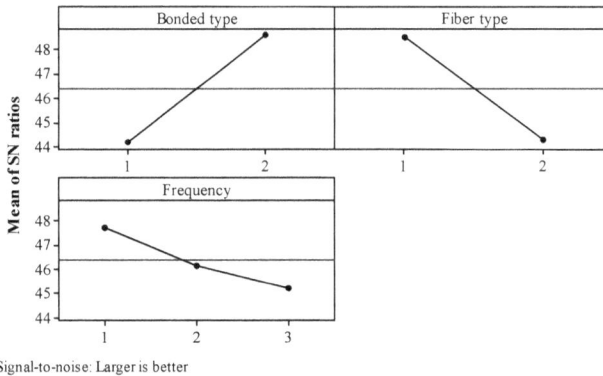

Signal-to-noise: Larger is better

Figure 32 Response graph for S/N ratio in first mode analysis

Figure 33 represents the average result of the voltage RMS recorded in the second mode experiment. All results exhibit a voltage range between 50mV to 500mV. The range is almost the same as that for first mode analysis. Generally, the embedded woven plate illustrates the lowest value, similar to first mode analysis, and displays almost constant results. Bonded woven and bonded unwoven plates demonstrate a decreasing trend with the increase of frequency. However, voltage RMS in embedded unwoven plates increased as the frequency increased from 80Hz to 100Hz and then decreased with the frequency increment.

The average results for the embedded woven plate are 87.4mV, 57.1mV, 64.6 and 112.3mV for 80Hz, 100Hz, 120Hz and 140Hz respectively. For the embedded unwoven plate, the average results are 147.2mV, 471.2mV, 301.8mV and 127.7mV for 80Hz, 100Hz, 120Hz and 140Hz respectively. The bonded woven average results are 391.9mV, 291mV, 218.8 and 170.9mV for 80Hz, 100Hz, 120Hz and 140Hz respectively. Meanwhile for bonded unwoven plates, the average results are 465.2mV, 257mV, 101.3 and 131.7mV for 80Hz, 100Hz, 120Hz and 140Hz respectively.

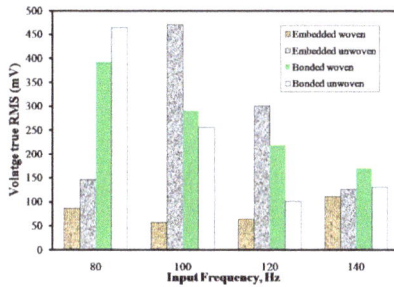

Figure 33 Voltage RMS results from the vibration plate in second mode analysis

Table 10shows the experimental parameters as a reference.

Table 11 and Figure 34 exhibit the response table and response graph of the S/N ratio for second mode analysis respectively. Following the analysis method established in the first mode, the fibre type is proposed as the most influential factor, followed by the bonding type and the frequency as shown in Table 11. Meanwhile, Figure 34 reports that the optimum parameters to achieve a higher voltage RMS is a combination of bonded MFC and unwoven kenaf fibre. The analysis suggested 120 Hz as the optimum frequency compared to the others. The natural frequency for bonded MFC on unwoven kenaf fibre from the modal testing is 125 Hz.

Table 10 The experimental parameters for second mode plate vibration testing analysis

Factor	Experimental Condition Levels			
	1	2	3	4
A-Bonding type of MFC patch	Embedded inside the Kenaf Plate	Bonded on the Kenaf Plate surface	0	0
B-Woven type of kenaf	Woven	Unwoven	0	0
C-Input frequency, Hz	80	100	120	140

Table 11 Response Table for S/N Ratio in second mode analysis

Level	Factors		
	Bonding type	Fibre type	Frequency
1	46.11	41.41	41.35
2	41.65	48.26	48.32
3	48.18		
4	43.4		
Delta	6.53	6.85	6.96
Rank	3	2	1

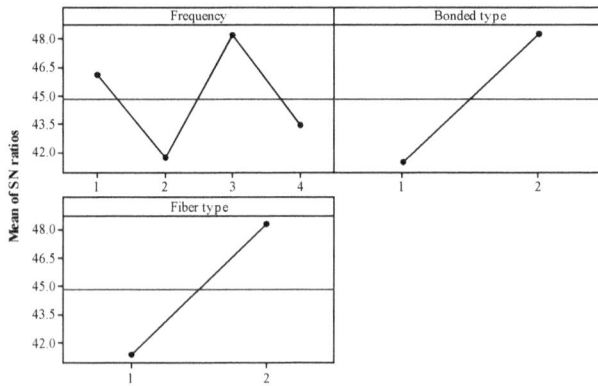

Signal-to-noise: Larger is better

Figure 34 Response graph for S/N ratio in second mode analysis

The correlation among the factors influencing the micro energy harvester is discussed in this section. First mode and second mode analysis have proposed different optimum parameters. In the first mode, the optimum parameter is a combination of bonded MFC and woven kenaf fibre and operates close to natural frequency. In the second mode, the optimum parameter to achieve a higher voltage RMS is a combination of bonded MFC, unwoven kenaf fibre and operates close to natural frequency. This shows that two different kenaf plates are proposed at different vibration range. The experiments show that the maximum voltage RMS occurred near to the natural frequency of the kenaf plate for the first and second mode.

Hence, the performance of the micro energy harvester for the kenaf plates is influenced by several factors including the fabrication technique, the weight of kenaf plate, the bonding MFC technique, the fibre type, the stiffness, and the excitation vibration. The bonded MFC technique may influence the damping percentage of the structure as reported in modal testing experiments. In addition, the fibre type (whether woven or unwoven) can affect the mechanical properties of the kenaf plate, hence affecting the stiffness of the structure.

Furthermore, the natural frequency of the kenaf plates presents a significant influence on the micro energy harvester performance. Excitation vibration that is near to the natural frequency causes the structures to deform at the maximum condition. Hence, a frequency which is close to the natural frequency is proposed for both the first mode and second mode conditions.

8. Conclusion

Generally, woven kenaf has shown better performance on mechanical properties compared to unwoven kenaf. The studies show that there was a difference of up to 199% for flexural strength, 178% for flexural modulus, 177% for tensile strength, and 84% for tensile modulus for woven kenaf compared to unwoven kenaf. The fiber orientation obviously influences the strength of kenaf polymer. Type A layers show a higher strength compared to type B layers. The density of yarn in each woven area increases the strength of the material. S1 (type A layer) shows higher strength compared to S4 (type B layer). The fracture in the composite is dominantly due to pull out, detaching and debonding mechanisms between fiber and matrix epoxy interface.

The comparison analysis indicates that type A woven orientation reports higher strength compared to R1, R2 and R3. The smaller size of the yarn and the higher density of yarn in each woven area improves the strength of kenaf polymer. The mechanical properties of woven kenaf may not achieve the quality of reinforced glass fiber reinforced. However, the effect of fiber orientation suggests a new idea in improving the mechanical properties of woven kenaf in the future application.

The modal testing is conducted on five kenaf plates. All plates demonstrated the same condition of deflection. The deflections occurred were bending, torsion, bending at different axis and combination of bending and torsion for the first mode, second mode, third mode and fourth modes respectively. The damping effect studied from analyses show that the bonded MFC type and deflection type may influence the damping percentage and trend for each plate. It shows that the damping percentage of both woven and unwoven kenaf plates increase at 100% and 50% respectively when bonded with MFC on their surfaces

Plate vibration testing suggests that the bonding technique is the most influential factor in micro energy harvesting at the vibration range of 20 to 60 Hz. Hence, MFC bonded to the surface is proposed as the best technique for promoting higher micro energy harvesting for woven kenaf plate at the vibration range of 20 to 60 Hz.

Several factors highlighted optimizing bonded technique including the damping effect due to MFC, the kenaf woven type, the distance from neutral axis, the stiffness of structure, the excitation vibration and the neutral frequency of a structure. This information may contribute as guidance for future designer and engineer. Besides that, the results can be used to analyze data in the research to identify the quality of micro energy generated in turbine blades.

Acknowledgements:

The authors would like to thanks Ministry of Education, Malaysia for providing Mybrain 15 grant (MyPhD) and Universiti Putra Malaysia for providing the research grants FRGS 5524501 and RUGS 978100 for this research work.

References

[1]   Akil, H. M., M. F. Omar, et al. (2011). "Kenaf fiber reinforced composites: A review." Materials and Design 32: 4107–4121.

https://doi.org/10.1016/j.matdes.2011.04.008

[2]   Ali, W. G. and S. W. Ibrahim (2012). "Power Analysis for Piezoelectric Energy Harvester." Energy and Power Engineering 4: 496-505.

https://doi.org/10.4236/epe.2012.46063

[3]   Anton, S. R. and H. A. Sodano (2007). "A review of power harvesting using piezoelectric materials (2003–2006)." Smart materials and structures 16: R1-R21.

https://doi.org/10.1088/0964-1726/16/3/R01

[4]   Antony, J. and M. Kaye (1999). Experimental quality—A strategic approach to achieve and improve quality. Massachusetts, Kluwer Academic Publishers.

[5]   Aziz, S. H. and M. P. Ansell (2004). "The effect of alkalization and fibre alignment on the mechanical and thermal properties of kenaf and hemp bast fibre composites: Part 1 - Polyester resin matrix." Compos Science Technology 64(9): 1219–30.

https://doi.org/10.1016/j.compscitech.2003.10.001

[6]   Azrin Hani, A. R., T. S. Chan, et al. (2013). "Impact and Flexural Properties of Imbalance Plain Woven Coir and Kenaf Composite." Applied Mechanics and Materials 271-272: 81-85.

https://doi.org/10.4028/www.scientific.net/AMM.271-272.81

[7]   Cao, Y., K. Goda, et al. (2007). Mechanical properties of kenaf fibers reinforced biodegradable composites. Proceedings of the 2007 International Conference on Advanced Fibers and Polymer Materials Vols 1 and 2: 299-302.

[8]   Daqaq, M. F., C. Stabler, et al. (2009). "Investigation of Power Harvesting via Parametric Excitations." Journal of Intelligent Material Systems and Structures 20(5): 545-557.

https://doi.org/10.1177/1045389X08100978

[9]   Daue, T. P., J. Kunzmann, et al. "Energy harvesting systems using piezo-electric macro fiber composites."

[10]   Davoodi, M. M., S. M. Sapuan, et al. (2010). "Mechanical properties of hybrid kenaf/glass reinforced epoxy composite for passenger car bumper beam." Materials and Design 31: 4927–4932.

https://doi.org/10.1016/j.matdes.2010.05.021

[11]   Erturk, A. and D. J. Inman (2009). "An experimentally validated bimorph cantilever model for piezoelectric energy harvesting from base excitations." Smart Materials and Structures 18(2).

https://doi.org/10.1088/0964-1726/18/2/025009

[12]   Faruk, O., A. K. Bledzkia, et al. (2012). "Biocomposites reinforced with natural fibers: 2000–2010." Progress in Polymer Science 37: 1552- 1596.

https://doi.org/10.1016/j.progpolymsci.2012.04.003

[13]   Hamdan, A., F. Mustapha, et al. (2014). "A review on the micro energy harvester in Structural Health Monitoring (SHM) of biocomposite material for Vertical Axis Wind Turbine (VAWT) system: A Malaysia perspective." Renewable and Sustainable Energy Reviews 35: 23-30.

https://doi.org/10.1016/j.rser.2014.03.050

[14]   Hani, A. R. A., R. Ahmad, et al. (2013). "Influence of Laminated Textile Structures on Mechanical Performance of NF-Epoxy Composites." International Scholarly and Scientific Research & Innovation 7(6): 757-763.

[15]   Hyun Jeong, S., Y.-T. Choi, et al. "Energy Harvesting Devices Using Macro-fiber Composite Materials." Journal of Intelligent Material Systems and Structures 21(6): 647-658.

https://doi.org/10.1177/1045389X10361633

[16]   Jenq, S. T., G. C. Hwang, et al. (1993). "The Effect of Square Cut-Outs on The Natural Frequencies and Mode Shapes of GRP Cross-Ply Laminates." Composite Science and Technology 47: 91-101.

https://doi.org/10.1016/0266-3538(93)90100-U

[17]   Larsen, G. C., M. H. Hansen, et al. (2002). Modal Analysis of Wind Turbine Blades. Risø National Laboratory, Roskilde, Denmark.

[18]   Lee, S. H. and S. Q. Wang (2006). "Biodegradable polymers/bamboo fiber biocomposite with bio-based coupling agent." Composites Part a-Applied Science and Manufacturing 37(1): 80-91.

https://doi.org/10.1016/j.compositesa.2005.04.015

[19]   Liu, W. J., M. Misra, et al. (2005). "'Green' composites from soy based plastic and pineapple leaf fiber: fabrication and properties evaluation." Polymer 46(8): 2710-2721.

https://doi.org/10.1016/j.polymer.2005.01.027

[20]   M.H. Kahrobaiyan, M. Asghari, et al. (2014). "A Timoshenko beam element based on the modified couple stress theory." International Journal of Mechanical Sciences 79: 78-83.

https://doi.org/10.1016/j.ijmecsci.2013.11.014

[21]   Masseran, N., A.M.Razali, et al. (2012). "An analysis of wind power density derived from several wind speed density functions: The regional assessment on wind power in Malaysia." Renewable and Sustainable Energy Reviews 16: 6476–6487.

https://doi.org/10.1016/j.rser.2012.03.073

[22]   Masseran, N., A. M. Razali, et al. (2012). "Evaluating the wind speed persistence for several wind stations in Peninsular Malaysia." Energy 37.

https://doi.org/10.1016/j.energy.2011.10.035

[23]   Nishino, T., K. Hirao, et al. (2003). "Kenaf reinforced biodegradable composite." Composites Science and Technology 63(9): 1281-1286.

https://doi.org/10.1016/S0266-3538(03)00099-X

[24]   Ochi, S. (2008). "Mechanical properties of kenaf fibers and kenaf/PLA composites." Mechanics of Materials 40(4-5): 446-452.

https://doi.org/10.1016/j.mechmat.2007.10.006

[25]   Oksman, K., M. Skrifvars, et al. (2003). "Natural fibres as reinforcement in polylactic acid (PLA) composites." Composites Science and Technology 63(9): 1317-1324.

https://doi.org/10.1016/S0266-3538(03)00103-9

[26]   Pabut, O., G. Allikas, et al. (2012). Model validation and structural analysis of a small wind turbine blade. 8th International DAAAM Baltic Conference, Tallinn, Estonia.

[27]   Park, R. and J. Jang (1997). "Stacking sequence effect of aramid – UHMPE hybrid composites by flexural test method." Polymer Testing 16: 549-562.

https://doi.org/10.1016/S0142-9418(97)00018-4

[28]   Park, S., J.-J. Lee, et al. (2008). "Electro-Mechanical Impedance-Based Wireless Structural Health Monitoring Using PCA-Data Compression and k-means Clustering Algorithms." Journal of Intelligent Material Systems and Structures 19(4): 509-520.

https://doi.org/10.1177/1045389X07077400

[29]   Phadke, M. S. (1989). Quality Engineering Using Robust Design. Englewood Cliffs, NJ, Prentice Hall.

[30]   Plackett, D., T. L. Andersen, et al. (2003). "Biodegradable composites based on L-polylactide and jute fibres." Composites Science and Technology 63(9): 1287-1296.

https://doi.org/10.1016/S0266-3538(03)00100-3

[31]   Ralib, A. A. M., A. Nurashikin, et al. (2009). Fabrication Techniques and Performance of Piezoelectric Energy Harvesters. 3rd International Conference on Energy and Environment, Malacca.

https://doi.org/10.1109/iceenviron.2009.5398605

[32]   Reza, M., M. Y. Jamaludin, et al. (2014). "Characteristics of continuous unidirectional kenaf fiber reinforced epoxy composites." Materials and Design 64: 640-649.

https://doi.org/10.1016/j.matdes.2014.08.010

[33]   Roslan, M. N., A. E. Ismail, et al. (2014). "Modelling Analysis on Mechanical Damage of Kenaf Reinforced Composite Plates under Oblique Impact Loadings." Applied Mechanics and Materials 465-466: 1324-1328.

https://doi.org/10.4028/www.scientific.net/AMM.465-466.1324

[34]   Shalwan, A. and B. F. Yousif (2013). "In State of Art: Mechanical and tribological behaviour of polymeric composites based on natural fibres." Materials & Design: 14-24.

https://doi.org/10.1016/j.matdes.2012.07.014

[35]   Shibata, S., Y. Cao, et al. (2008). "Flexural modulus of the unidirectional and random composites made from biodegradable resin and bamboo and kenaf fibers." Composites Part A: Applied Science and Manufacturing 39: 640-6.

https://doi.org/10.1016/j.compositesa.2007.10.021

[36]   Sodano, H. A., D. J. Inman, et al. (2005). "Comparison of piezoelectric energy harvesting devices for recharging batteries." Journal of Intelligent Material Systems and Structures 16(10): 799-807.

https://doi.org/10.1177/1045389X05056681

[37]   Song, H. J., Y.-T. Choi, et al. (2009). "Performance Evaluation of Multi-tier Energy Harvesters Using Macro-fiber Composite Patches." Journal of Intelligent Material Systems and Structures 20(17): 2077-2088.

https://doi.org/10.1177/1045389X09347017

[38]   Sopian, K., M. Y. H. Othman, et al. (1995). "The wind energy potential of Malaysia." Renewable Energy 6(8): 1005-1016.

https://doi.org/10.1016/0960-1481(95)00004-8

[39]   Stuart, T., Q. Liu, et al. (2006). "Structural biocomposites from flax - Part I: Effect of bio-technical fibre modification on composite properties." Composites Part a-Applied Science and Manufacturing 37(3): 393-404.

https://doi.org/10.1016/j.compositesa.2005.06.002

[40]   Taguchi, G. (1990). Introduction to Quality Engineering. Tokyo, Asian Productivity Organization.

[41]   Tien, C. M. T. and N. S. Goo (2010). "Use of a piezo-composite generating element for harvesting wind energy in an urban region." Aircraft Engineering and Aerospace Technology: An International Journal 82: 376-381.

https://doi.org/10.1108/00022661011104538

[42]   Wambua, P., J. Ivens, et al. (2003). "Natural fibres: can they replace glass in fibre reinforced plastics?" Composites Science and Technology 63: 1259–1264.

https://doi.org/10.1016/S0266-3538(03)00096-4

[43]   Wang, W., Z. Yang, et al. (2012). Macro fiber composite (MFC) piezoelectric actuator excitation method for in flight leading edge deicing. Advances in Structural Health Management and Composite Structures 2012 (ASHMCS 2012), Jeonju, Republic of Korea.

[44]   Yahaya, R., S. Sapuan, et al. (2014). "Mechanical performance of woven kenaf-Kevlar hybrid composites." Journal of Reinforced Plastics and Composites 33(24): 2242-2254.

https://doi.org/10.1177/0731684414559864

[45]   Yahaya, R., S. M. Sapuan, et al. (2014). "Effects of Kenaf Contents and Fiber Orientation on Physical, Mechanical, and Morphological Properties of Hybrid Laminated Composites for Vehicle Spall Liners." Polymer composites: 1469–1476.

[46]   Yahaya, R., S. M. Sapuan, et al. (2014). "Mechanical performance of woven kenaf-Kevlar hybrid composites." Journal of Reinforced Plastics and Composites.

https://doi.org/10.1177/0731684414559864

[47] Yahaya, R., S. M. Sapuan, et al. (2015). "Effect of layering sequence and chemical treatment on the mechanical properties of woven kenaf–aramid hybrid laminated composites." Materials and Design 67: 173-179.

https://doi.org/10.1016/j.matdes.2014.11.024

[48] Yousif, B. F., A. Shalwan, et al. (2012). "Flexural properties of treated and untreated kenaf/epoxy composites." Materials and Design 40: 378-385.

https://doi.org/10.1016/j.matdes.2012.04.017

[49] Yousif, B. F., A. Shalwan, et al. (2012). "Flexural properties of treated and untreated kenaf/epoxy composites." Materials and Design 40: 378-385.

https://doi.org/10.1016/j.matdes.2012.04.017

[50] Zampaloni, M., F. Pourboghrat, et al. (2007). "Kenaf natural fiber reinforced polypropylene composites: A discussion on manufacturing problems and solutions." Composites: Part A 38: 1569-1580.

https://doi.org/10.1016/j.compositesa.2007.01.001

Chapter 6

The macro fiber composite (MFC) bonded effect analysis on the micro energy harvester performance and structural health monitoring system of woven kenaf turbine blade for vertical axis wind turbine application

A. Hamdan[1], F. Mustapha[2]

[1]Department of Aerospace Engineering, Universiti Putra Malaysia, 43400 Serdang, Selangor, Malaysia,

[2]Aerospace Manufacturing Research Centre (AMRC), Level 7, Tower Block, Faculty of Engineering, Universiti Putra Malaysia, 43400 Serdang, Selangor, Malaysia

Keywords

Smart Polymers, Biocomposite, Taguchi Method, Woven Kenaf, Structural Health Monitoring (SHM)

Abstract

The application of Vertical Axis Wind Turbine (VAWT) is suitable for a low wind speed environment. Nevertheless, VAWT with kenaf turbine blade will promote additional green concept by utilizing biocomposite material. The innovation in turbine blade via Macro Fiber Composite (MFC) as Structural Health Monitoring (SHM) system and micro energy harvester can enhance the VAWT technology application. Hence, this research objective is to evaluate the factors influencing the performance of micro energy harvester and to assess the feasibility of SHM application for biocomposite turbine blades. There are two methods to attach the MFC used in this study which are surface bonded and embedding into the turbine blade. Vibration simulation experiment and modal testing approach are conducted on the kenaf turbine blade and further analysis performed via Taguchi statistical analysis to determine the factors affecting the micro energy harvester and SHM performance. The results show that bonded to the surface is proposed as the best technique for promoting higher micro energy harvesting at the vibration range of 10 to 90 Hz. Furthermore, SHM system is proven to operate simultaneously with micro energy harvesting system in turbine blades.

Contents

1. Introduction

Wind turbines can be categorised into two main types, depending on the axial direction of the rotor shaft. One type is the Horizontal Axis Wind Turbine (HAWT) and the second type is the Vertical Axis Wind Turbine (VAWT). HAWTs have blades mounted radially from the rotor. Modern types usually have two or three blades and are generally used for large scale grid electrical power generation. VAWTs are not as common and have only recently been used for large scale electricity generation. Both types of wind turbine are being rigorously tested and improved (Herbert, Iniyan et al. 2007).Several studies show that the application of the VAWT has more advantages compared to the HAWT (Sandra

Eriksson, Bernhoff et al. 2008; Dabiri 2011; Aslam Bhutta, Hayat et al. 2012; Mohamed 2012). A comparison between the VAWT and the HAWT is presented in Table 1 (Aslam Bhutta, Hayat et al. 2012). The VAWT does not have to be orientated to the wind direction. Also, it does not need a tower hence reducing the capital cost. In fact, the generator is mounted at ground level for easy access (Kanellos and Hatziargyriou 2008; Yeh and Wang 2008; Ibrahim 2009). Additionally, recent studies show that VAWTs can be installed much closer to each other compared to HAWTs, so the power density per square meter could be considerably higher than for the configurations used presently (Dabiri 2011).

For various reasons, there is now a resurgence of interest in VAWTs, in particular Darrieus turbines (Mohamed 2012). Furthermore, VAWTs exhibit more advantages compared to HAWTs in terms of Malaysia's weather conditions and risks. They harmonise with the circumstances such as low average wind velocity, lightning risk and bird strike risk. The types of VAWT are further analysed and reviewed in the next paragraph.

Table 1 Comparison between VAWTs and HAWTs (Aslam Bhutta, Hayat et al. 2012)

	Vertical axis wind turbine (VAWT)	Horizontal axis wind turbine (HAWT)
Tower sway	Small	Large
Yaw mechanism	No	Yes
Self starting	No	Yes
Overall formation	Simple	Complex
Generator location	On ground	Not On ground
Height from ground	Small	Large
Blade's operation space	Small	Large
Noise produced	Less	Relatively high
Wind direction	Independent	Dependent
Obstruction for birds	Less	High
Ideal efficiency	More than 70%	50–60%
Wind velocity for start	Very low	Relatively high

The study of VAWT configuration has already been conducted and is established. There are several configurations listed, which are as follows: the Darrieus rotor – egg beater shaped (Figure 1), the Darrieus rotor–straight bladed (Figure 3), the Darrieus rotor–variable geometry oval trajectory (VGOT) (Figure 2), the Darrieus–Masgrowe, the twisted three bladed Darrieus rotor, the Crossflex, Savonius rotor (Figure 4), the Combined Savonius and Darrieus rotor, the two leaf semi rotary, Sistan wind mill and the Zephyr turbine (Aslam Bhutta, Hayat et al. 2012). In terms of the manufacturing process and fabrication costs, the Darrieus rotor-straight blade or giromill showed a reliable configuration. The two blades of the giromill is generally named the H-rotor (Mertens, van-Kuik et al. 2003; Howell, Qin et al. 2010). Hence, a deep consideration and review will be conducted on the Darrieus rotor-straight blade. In terms of the performance of the VAWT, the central shaft produces a higher impact to the vibration (Li 2012). Besides this, the efficiency of the VAWT is increased by the greater length of the VAWT and the diameter of the turbine blades (Li 2012).

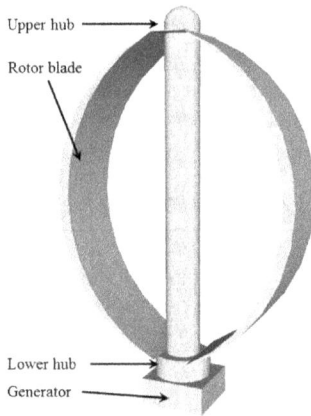

Figure 1 Darrieus rotor – egg beater shaped (Aslam Bhutta, Hayat et al. 2012)

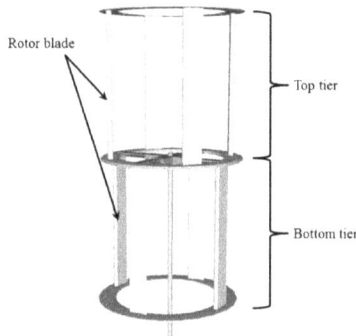

Figure 2 Darrieus rotor –VGOT (Aslam Bhutta, Hayat et al. 2012)

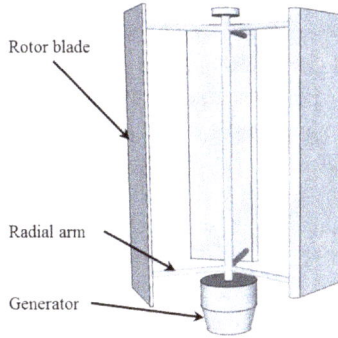

Figure 3 Darrieus rotor –straight bladed (Aslam Bhutta, Hayat et al. 2012)

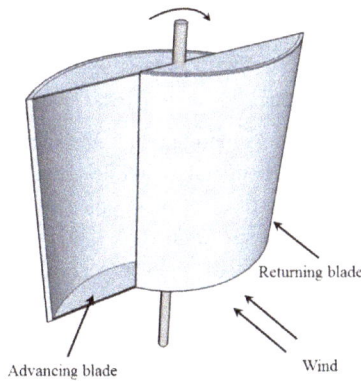

Figure 4 Savanius rotor (Aslam Bhutta, Hayat et al. 2012)

Previous study on VAWT performance has focused on torque (Gavald, Massons et al. 1990; Islam, Ting et al. 2008; Chong, Poh et al. 2012; Mohamed 2012; Park, Asim et al. 2012), power (Gavald, Massons et al. 1990; Hossain, Iqbal et al. 2007; Chong, Poh et al. 2012; Greenblatt, Schulman et al. 2012; Mohamed 2012; Park, Asim et al. 2012) and rotational speed (Chong, Poh et al. 2012). Several factors influencing VAWT performance were discussed, such as blade health, fluid flow around the blade, and wind turbine design. The studies were conducted using simulation and experimental methods and hence the VAWT performance was improved.

In the design analysis, the design configuration may include a combination of several existing VAWT types (Gavald, Massons et al. 1990). A new design of the Darrieus and Savonius combined rotor is proposed and analysed. The starting torque shows an improvement. In addition, a comparison of existing standard airfoil shapes was conducted for 20 designs (Mohamed 2012). From the 20 designs suggested and subjected to computational analysis, the H-rotor Darrieus turbine (involving the S-1046 type of airfoil) appears to be very suitable for wind energy generation, particularly in urban areas. Greenblattet et. al. (2012) proposed a plasma actuator for controlling the flow separation, hence increasing the power generated by about 38%. Fixed and variable pitch blades were also studied. The results show that variable pitch blades manage to overcome the starting torque issues associated with VAWTs.

Furthermore, additional accessory may help in improving the amount of power generated. The guide vane at the outer devices of the VAWT system may act as the Bernoulli principle. The reductions in air pressure causes the air to flow into the tunnel at a higher velocity compared to the outer velocity. This improves the rotational speed and the starting behaviour performance (Takao, Kuma et al. 2009; Chong, Naghavi et al. 2011; Chong, Fazlizan et al. 2012; Chong, Poh et al. 2012; Chong, Pan et al. 2013).

The experimental output was also supported and enhanced by simulation analysis. The research on vortex simulation, dynamic stall and height-to-diameter ratio shows a better understanding and explanation of the aerodynamic problem in the experiment (Vandenberghe and Dick 1987; Islam, Amin et al. 2007; Sukanta, Agnimitra et al. 2010; Stein, Hsu et al. 2012). The selection of codes in computational fluid dynamics analysis is also optimised and improved in the simulation study. Hence, an accurate result could be projected (Tai, Kang et al. 2012). An analysis of blade health was conducted as well. It studies the effect of a faulty blade on the torque and power output. It shows that the torque and power could decrease as the number of missing blades increases (Park, Asim et al. 2012).

From an economic point of view, one study showed that the improvement in aerodynamic design could benefit by about 6 cent for each kW/h generated by the designed VAWT (Saeidi, Sedaghat et al. 2013). A summary of the above discussion is presented in Table 2. Both simulation and experimental methods were conducted to improve the technology of VAWTs, especially for the RE fields. The design analysis seems to have reached maturity. Furthermore, the researchers have paid significant attention to the parameters involved, such as aerodynamic performance and fluid flow analysis. However, there are still research opportunities to be pursued in optimising blade design for several factors, such as span length, chord length, manufacturing ability and aerodynamic shape. Further analysis can also be conducted to study those factors which are most influential in VAWT performance. However, the studies on structural integrity need to be explored further; the issues highlighted in structural integrity involve: the structure's critical point, blade vibration utilisation, the effect of the dynamics of wind flow in a very short period, the structural health monitoring system, the natural frequency of the structure and material selection.

Therefore, the application of Structural Health Monitoring (SHM) system in turbine blade is becoming very important. There are several methods available for detecting damage or failure in the blade structure via SHM: Acoustic Emission, Ultrasonic, Fibre Optic, Laser Doppler Vibrometer and Thermal Imaging (Ciang, Lee et al. 2008). Additionally, a new smart material (piezoelectric material) offers a better technology, which not only detects damage but acts as an actuator and harvests electricity as well (Anton and Sodano 2007; Liu, Zhang et al. 2012). In this system, the sensor will react if several situations occur, such as a curvature of the structure, a strain state in the sensors, damage in the structure and a failure mode of the structure called buckling (Sundaresan, Schulz et al. August 1999). The active composite fibre sensor acts as well as SHM as it offers the following advantages: low weight, low cost compared to other devices, and is easy to install (Sundaresan and Schulz August 2006).

The application of piezoelectric material as an enhancement of the VAWT in energy harvesters could become a new challenge. As the VAWT is the main energy collector, piezoelectric material can act as a micro energy harvester or wind vibration energy harvester (Liu, Zhang et al. 2012). The technology is focused to become an alternative to the conventional battery, or proposed as a lifelong battery for low power subsystems or devices (Sodano 2003; Liu, Zhang et al. 2012).

Table 2 Summary of research conducted on the Darrieus rotor-straight blade of the VAWT

No	Author	Performance parameter	Design parameter	Improvement
1	DavoodSaeidi, Ahmad Sedaghat, PouryaAlamdari, Ali Akbar Alemrajabi(Saeidi, Sedaghat et al. 2013)	Aerodynamic design and economical evaluation	Aerodynamic performance, Power production	The evaluation shows a profit of 6 cent per each kW/h generated power by the designed VAWT.

No	Author	Performance parameter	Design parameter	Improvement
2	Kyoo-seon Park TaimoorAsim Rakesh Mishra (Park, Asim et al. 2012)	The torque and power outputs from the VAWT	Blade health	As the number of missing blades increases, the torque and power outputs from the VAWT decrease.
3	M.H. Mohamed (Mohamed 2012)	To maximise output torque coefficient and output power coefficient(efficiency)	Aerodynamic investigation Airfoil design Mutual interaction between blades (solidity effect)	The optimal configuration of the H-rotor Darrieus turbine involving S-1046 appears to be very promising for wind energy generation, in particular in urban areas.
4	David Greenblatt, Magen Schulman, Amos Ben-Harav (Greenblatt, Schulman et al. 2012)	Turbine power	Dynamic flow Separation control using plasma actuators Viability of up-scaling the turbine. Baseline turbine and the effect of slip-rings	Turbine power of up to 38% were measured, up-scaling the turbine by a factor of 5 and 10.
5	Feng-Zhu Tai, Ki-Weon Kang, Mi-Hye Jang, Young-Jin Woo, Jang-Ho Lee(Tai, Kang et al. 2012)	LDWT code of Darrieus half egg beater shape	Tip speed ratio (TSR) Reynolds number	LDWT codes show a better match with the test data in the higher TSR region than DART code, previously researched by the Sandia National Laboratory.
6	W.T. Chong, A. Fazlizan, S.C. Poh, K.C. Pan, H.W. Ping (Chong, Fazlizan et al. 2012)	Starting behaviour and rotational speed	A novel power-augmentation-guide-vane (PAGV)	Power-augmentation-guide-vane (PAGV) increased the speed of the high altitude free stream wind for optimum wind energy extraction. Hence this increased the rotational speed.

No	Author	Performance parameter	Design parameter	Improvement
7	Jna. Gavald, J. Massons, F. Diaz (Gavald, Massons et al. 1990)	Starting torque & power coefficient	New design	New design of the Darrieus & Savonius combined rotor is proposed and analysed.
8	D. Vandenberghe, E. Dick (Vandenberghe and Dick 1987) Islam, M., Amin, M.R., Ting, D.S.K., Fartaj, A. (Islam, Amin et al. 2007) Sukanta, Roy.,Agnimitra, Biswas., Rajat, Gupta(Sukanta, Agnimitra et al. 2010)	Comparison between experimental and numerical analysis	Vortex simulation of the wake and the modelling of dynamic stall (Chong, Naghavi et al. 2011; Chong, Poh et al. 2012) Power coefficient (Cp) at different height-to-diameter (H/D) ratios (Chong, Naghavi et al. 2011)	Carried out a detailed aerodynamic study of this type of configuration. Closed comparison between experimental and numerical analysis.
9	Takao, M.,Kuma, H.,Maeda, T.,Kamada, Y.,Oki, M.,Minoda, A. (Takao, Kuma et al. 2009)	Aerodynamic performance	Power coefficient	Performance parameters improved by addition of guide vane row around the turbine. Power coefficient is 0.215, which is 1.8 times higher than that of the original turbine without any guide.
10	Chong, W.T, Poh, S.C, Fazlizan, A.Pan, K.C (Chong, Poh et al. 2012)	Rotational speed, torque output and power output	Omni-directional guide vane	Omni-directional guide vane increases the rotational speed to two times better than the original. Increases torque output by 206% for tip speed ratio of 0.4.

No	Author	Performance parameter	Design parameter	Improvement
11	P. Stein, M.-C. Hsu, Y. Bazilevs, K. Beucke (Stein, Hsu et al. 2012)	Aerodynamic performance	Comparison data	Preliminary aerodynamic simulation of a newly constructed VAWT model in 3D under realistic wind conditions and rotation speed is presented.
12	M. Islam, D.S.K. Ting, A. Fartaj (Islam, Ting et al. 2008)	Starting torque	Variable blades	Variable pitch blades have the potential to overcome the starting torque issues associated with VAWTs
13	Chong, W.T., Naghavi, M.S., Poh, S.C., Mahlia,T.M.I., Pan, K.C. (Chong, Naghavi et al. 2011)	Estimated annual energy saving	A novel power-augmentation-guide-vane (PAGV)	Estimated energy saving for wind mill system with the PAGV and an H-rotor VAWT mounted on the top of a 220 m high building is 195.2 MW h/year.
14	Chong, W.T., Pan, K.C.,Poh, S.C., Fazlizan, A.,Oon, C.S., Badarudin, A.,Nik-Ghazali, N. (Chong, Pan et al. 2013)	Rotational speed, power output, rotor torque	A novel power-augmentation-guide-vane (PAGV)	The power output increment of the rotor is 5.8times, the wind rotor rotational speed is increased by 75.16% and simulation study on the rotor torque is increased by 2.88 times with the PAGV
15	John O. Dabiri (Dabiri 2011)	Power density	Counter rotating vertical-axis wind turbine arrays	It increases the power density compared to the HAWT and is capable of alleviating many of the practical challenges associated with large HAWTs, such as the cost and logistics of their manufacture, transportation and installation.

This concept was applied in unmanned aerial vehicle (UAV) technology (Anton and Inman 2008) and shows promising results. It will be based on mechanical vibration, mechanical stress and strain, thermal energy from furnaces, heaters and friction sources, sunlight or room light, the human body, chemical or biological sources, which can generate mW or μW level power (Heung, Joo-Hyong et al. 2011). In the VAWT, mechanical vibration is the main factor generating power in piezoelectric material.

The research studies on piezoelectric material are reviewed for VAWT application. Flynn and Sander (Flynn and Sanders 2002) imposed fundamental limitations on lead zirconate titanate (PZT) material and indicated that the mechanical stress limit is the effective constraint in typical PZT materials. They reported that a mechanical stress-limited work cycle was 330W/cm^{3}at 100 kHz for PZT-5H. On the other hand, a study of different types of piezoelectric material was conducted in Inman's group such as monolithic piezoelectric ceramic, bimorph Quick Pack (QP) actuator and MFC (Sodano, Park et al. 2003; Sodano, Inman et al. 2004; Sodano, Inman et al. 2005; Erturk, Bilgen et al. 2008; Erturk and Inman 2009). The papers report on the efficiency of MFC and PZT and the capacity to recharge a discharged battery of the three types of material. Furthermore, a PZT piezoelectric cantilever is proposed as a micro machined Si proof mass in a low frequency vibration application (Shen, Park et al. 2009).

Preliminary research has been conducted on a complete autonomous sensing unit that incorporates SHM and power harvesting technologies into a single, self-powered device (Anton and Sodano 2007). Anton et al. (Anton, Taylor et al. 2012) report that a multisource energy harvester is capable of powering a wireless impedance device sensor node through solar and vibration energy harvesting. The experiment was conducted on a full scale wind turbine during bright, cloudy, mild wind and high wind conditions. It sufficiently charged its input capacitor (0.1 F) to 3.6 V in an acceptable period time.

Several studies of material effectiveness have been carried out by other researchers, involving such materials as piezoelectric ceramics with a microstructure texture containing a template of SrTiO3 (STO) (Jeong, Lee et al. 2011), multilayer piezocomposite composed of layers of carbon/epoxy, PZT ceramic and glass/epoxy (Tien and Goo 2010) (Figure 5), piezoelectric polymer polyvinylidene fluoride (PVDF) (Farinholt, Pedrazas et al. 2009), ionically conductive ionic polymer transducer (Farinholt, Pedrazas et al. 2009) and aluminum nitride as a piezoelectric material (Elfrink, Kamel et al. 2009). Tien and Goo (Tien and Goo 2010) reported that PZT performance depends on the distance of the PZT layer from the neutral axis of the structure as shown in Figure 5. This situation is related to the Bending Theory.

In addition, several researchers have also proposed other methods in order to increase output power effectiveness, such as optimising power conditioning circuitry (Anton and Sodano 2007), using different beam shapes (Goldschmidtboeing and Woias 2008) and using multilayer structures (Zhu, Almusallam et al. 2010). In term of material selection, MFC is suggested due to some of its advanced features. Its flexibility makes it very suitable for bonding on large and vibrating structures and also its high electromechanical coupling coefficient (Hyun Jeong, Choi et al. 2010).

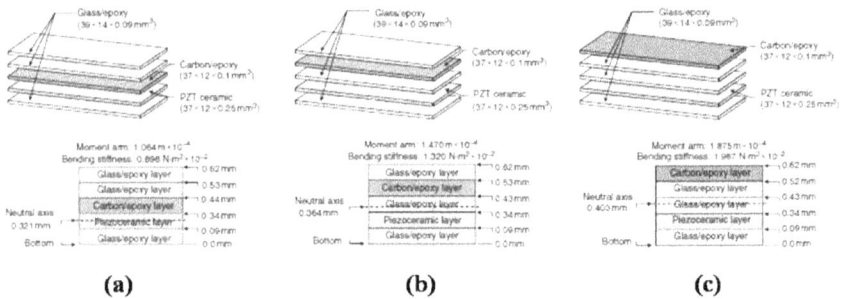

(a)   (b)   (c)

Figure 5 Geometry of multilayer piezocomposite(Tien and Goo 2010)

The discussions show that SHM for VAWTs can be enhanced to include application in the micro energy harvester. MFC is proposed for this due to its various advantages. Furthermore, previous studies have shown that most of the research has focused on the vibration cantilever beam. However, studies on the self energy of VAWT need to be undertaken and experimentally validated in the current Malaysian environment and regarding user friendly criteria. Several aspects, such as the MFC bonding technique, the natural frequency of the structure and the material of the turbine blades, will be highlighted.

2.   Experimental setup

2.1   Wind turbine fabrication

The wind turbine blade is fabricated via vacuum infusion process. The mould is fabricated in the laboratory following the characteristics of NACA 0018 design specification. The complete mould is shown in Figure 6. A woven kenaf which is 3 X 11 yarn for each 1 cm^2 utilizing 300 tex yarn is selected as shown in Figure 7. The woven kenaf is stacked in three layers as shown in Figure 8.

Figure 6 Top and bottom mould of turbine blade

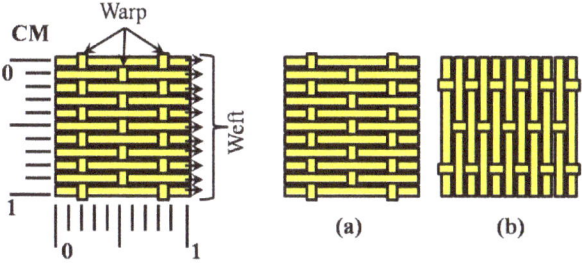

Figure 7 Schematic of kenaf woven orientation (a) Orientation A, (b) Orientation B

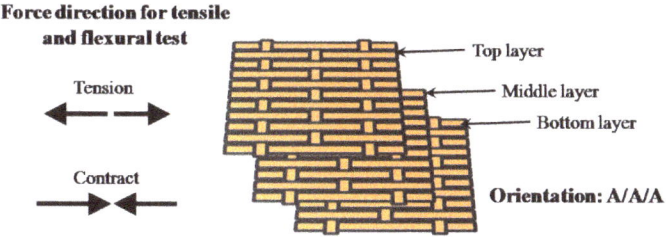

Figure 8 Sample 1: Orientation A/A/A

There are two techniques used to bond the MFC patch. For the first technique, the MFC patch is embedded inside the turbine blade. The other technique is onto the turbine blade where the patch is bonded on the surface by assuming that the bonding is effectively adhered so that the stress is effectively transferred between two surfaces.

In the fabrication process, the ratio between kenaf and epoxy resin is 30:70 and it is cured for 12 hours. The epoxy employed in this research is EpoxAmite 100 (Smooth-on) and 102 Medium Hardener (Smooth-on). The resin was prepared at a combination of epoxy to hardener, 100g: 28.4g as recommended by the manufacturer. In order to reduce cost, reduce weight and manufacturing process time, the blades were fabricated in half design and become one plate as shown in Figure 9. After cured, the part was taking out and cut to the required dimension. MFC patch was embedded approximately near to center line of turbine length and width as shown in Figure 9. In this research, the location was selected for the fabrication and assembly convenient. The research on MFC location optimization can be conducted in the future. The MFC was properly layed between the kenaf layer as shown in plate fabrication.

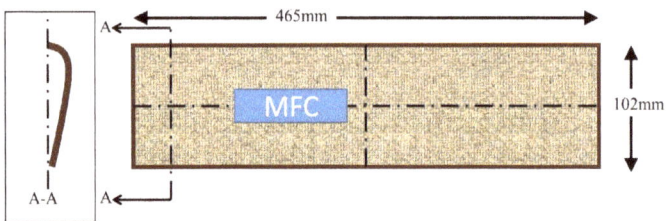

Figure 9 The schematic picture turbine blade design and MFC location

The MFC was embedded and bonded at the same location of the woven kenaf turbine blade. After curing process, a second MFC was bonded at the same place of embedded MFC located as shown in Figure 10. Hence, the data collection could be compared thoroughly. Four kenaf blades were fabricated for assembly in the wind turbine system prototype as shown in Figure 11. The blade containing the MFC patch was employed for micro energy harvesting in wind turbines.

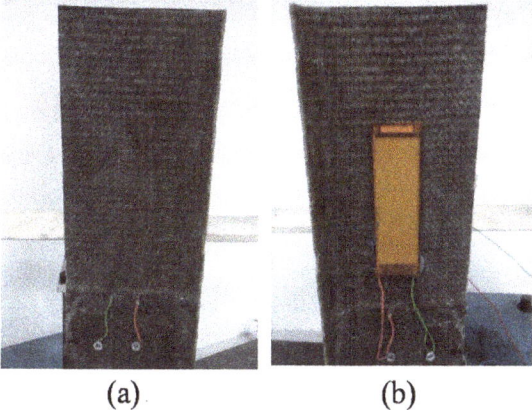

Figure 10 Woven kenaf turbine blade (a) embedded MFC location, (b) Bonded MFC located at same place of embedded MFC

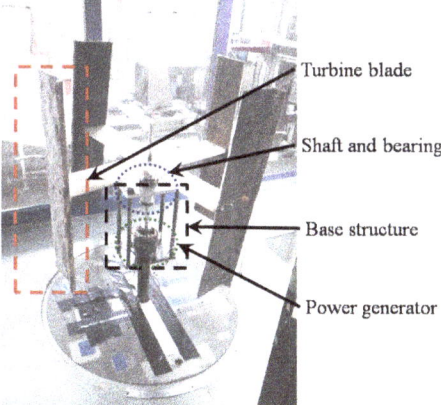

Figure 11 Segment in wind turbine system

2.2 Micro energy harvester testing

A vibration test for the turbine blades was conducted on the wind turbine prototype. The wind turbine was fixed and the respective turbine blade was exerted by a shaker at frequency range from 10Hz to 100Hz. The range was proposed by assuming that a first mode was occurring in that range. This test only gives attention to first mode of turbine blade. The same methodology as for the vibration test on the plates was employed for this testing. The experiments were repeated 6 times and each experiment conducted in 10 seconds. The MFC is randomly located in layer as this research is focussed to determine the bonding MFC technique effect and material optimization. Experimental setup is shown in Figure 12. Further investigation on the MFC is conducted for Structural Health Monitoring (SHM) application.

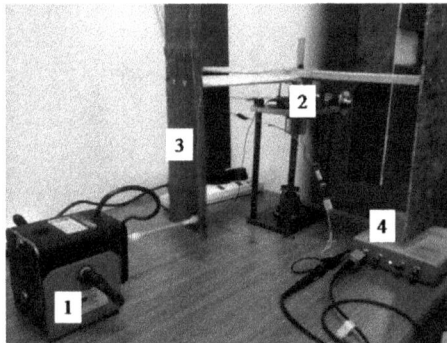

Legend
1. Shaker   3. Turbine blade
2. Wind turbine 4. Oscilloscope

Figure 12 Experimental setup for micro energy harvesting testing on wind turbine system prototype

2.3 Structural health monitoring (SHM) identification for turbine blade via modal approach

The Structural Health Monitoring (SHM) test is conducted to determine the pattern of frequency response function graph between embedded and bonded MFC as well as between damaged and undamaged turbine blades. This experiment was conducted as a preliminary study on a SHM system on kenaf turbine blades, since the main focus of this research is on micro energy harvesters. In order to simulate damage turbine blade, the blade consisting of a MFC patch was drilled at the tip blade as shown in Figure 13.

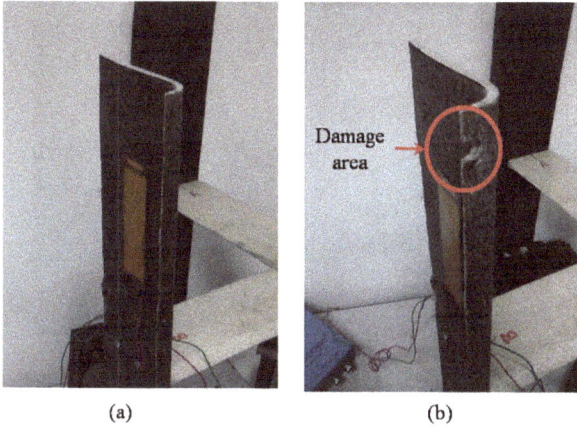

Figure 13 Turbine blade (a) normal condition, (b) damage condition

The experiment equipment is prepared into two sections which are vibration exciter and data collector. The vibration exciter includes equipment, shaker, amplifier and vibration controller. Whilst for the data collector, MFC as sensor and oscilloscope were employed. The integration between vibration exciter, data collector and wind turbine system were set up as shown in Figure 14. The tip of the shaker was touched to the free end of the plate to simulate the vibration at respective ranges of frequency.

The SHM identification is setup in following the modal approach and vibration concept as conducted by Sodano, Wang and Ali (Sodano, Inman et al. 2005; Ali and Ibrahim 2012; Wang 2012). As the objective is only to determine the pattern graph of frequency response function, the analysis is only limited to data comparison between embedded and bonded MFC as well as damaged and undamaged turbine blade.

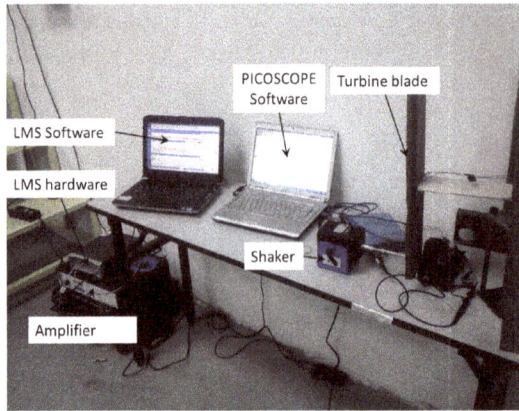

(a)

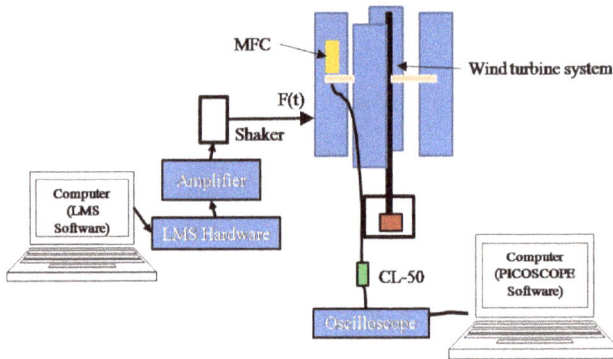

(b)

Figure 14 Experimental setup for structural health monitoring system identification (a) The arrangement of equipment (b) The Schematic detail of experiment setup

The shaker vibrations were set at the range of between 10 Hz to 100 Hz. The amplifier was set at 2 watt. The shaker frequency and amplitude was controlled by a frequency controller and amplifier respectively. The shaker frequency was set at 10, 20, 30, 40, 50, 60, 70, 80, and 90Hz. It operates at a bandwidth of 100 Hz, spectral line equal to 1024 at 0.097Hz resolution and an acquisition time of 10.24s. The shaker vibration is set to excite in burst sine signal.

The experiment started with the normal turbine blade for bonded and embedded MFC. Then, the turbine blade was drilled and the procedure was repeated for the damage turbine blades in bonded and embedded MFC condition. The voltage generated induced from MFC was transmitted to a smart power harvesting module (EH-CL 50 from smart-material) before connect to oscilloscope for voltage recording. The spectrum data was set similar to the shaker spectral line which is 1024. The voltage RMS was recorded via oscilloscope (PICOSCOPE) and analyzed with the PICOSCOPE software. It displays a frequency response function graph by employing the flat top window function. The flat top windows function is calculated with the data from the time domain function to frequency response function. The Flat top function offers low errors compare to other functions. Logarithmic unit: dBV was employed for describing the magnitude of each frequency. Further analysis via the Taguchi method is employed from the experimental data.

2.4 Statistical analysis via Taguchi method

The Taguchi methods are statistical methods initially developed by Genichi Taguchi to improve the quality of manufactured goods. More recently, the techniques are used in scientific and engineering experiments since they allow for the analysis of many different parameters without a prohibitively high amount of experiments. Many researchers nowadays apply robust design as a tool to achieve quality engineering in many fields.

Furthermore, Taguchi offers an experimental design totally based on statistical design as a tool which is less sensitive to noise factors. Two major tools used are: signal to noise (S/N) ratio, which measures quality with emphasis on variation, and orthogonal arrays, which accommodate many design factors simultaneously. The significance of the factor or multiple factors that affect the quality and performance can be determined in a very short time when this technique is employed.

The method of calculating the S/N ratio response is designed in three different modes depending on whether the quality characteristics is smaller the better, larger the better or nominal the better (Taguchi 1990). In this analysis, the larger the better is preferred to perform high energy. The equations for calculating the S/N ratio are for the larger the better characteristic (in dB) which is

$$S/N = -10 \log \frac{1}{n} \left(\sum \frac{1}{y_i^2} \right).$$

n is the number of observations and y_i is the observed data. The S/N ratio values function as a performance measurement to develop processes insensitive to noise factors.

The degree of predictable performance of a product or process in the presence of noise factors could be defined from the S/N ratio values. For each type of characteristics, with the above S/N ratio, the higher the S/N ratio, the better the result. The S/N ratio was presented in a response graph and response table.

In the micro energy harvesting analysis on turbine blade prototypes, experiments were conducted at L_{14} orthogonal array. Two factors employed with two levels and three levels in each factor as shown in Table 3. Voltage RMS was the output for this analysis.

Table 3 Taguchi experimental parameter for micro energy harvesting analysis on turbine blade prototype

Factor	Level 1	Level 2	Level 3
Bonded type	Embedded	Bondded	
Frequency, Hz	20	40	60

3. Result and discussion

Figure 15 shows the voltage RMS result on micro energy harvesting from turbine blade prototypes. Bonded MFC demonstrate higher value as compared to embedded MFC. The average result for bonded MFC in the range from 10Hz to 100Hz is 117.3mV. Meanwhile, the average result for embedded MFC is 26.2mV. Bonded MFC shows a 348% increment compared to embedded MFC. This shows that the bonded MFC performs better than embedded MFC in harvesting energy. The results are further analyzed via a Taguchi statistical analysis.

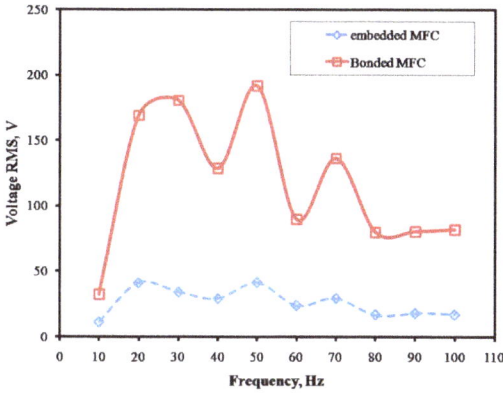

Figure 15 Voltage RMS result in micro energy harvesting on prototype of wind turbine

4.   Taguchi statistical analysis for micro energy harvesting of turbine blade prototypes

Statistical analysis on micro energy harvesting of turbine blade prototype is presented via response table and response graph in Table 4 and Figure 16 respectively. Table 4 shows that the bonded type is the most significant factor influencing the results of voltage RMS in turbine blades. Meanwhile, Figure 16 exhibits that the optimum parameter to achieve higher voltage RMS is a combination of MFC bonded to turbine blade.

Table 4 Response Table for S/N ratio in micro energy harvested of turbine blade prototype

Level	Factors	
	Bonded type	frequency
1	29.74	38.3
2	41.83	35.67
3		33.38
Delta	12.09	4.91
Rank	1	2

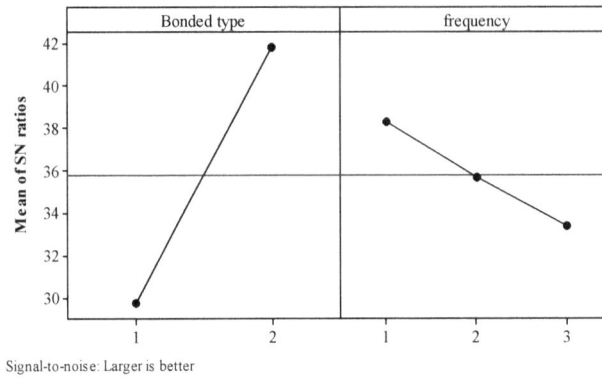

Figure 16 Response graph for S/N ratio in micro energy harvested of turbine blade prototype

The results show a comparable finding with plate vibration testing which proposed bonded MFC as the best parameter. Figure 16 exhibits that frequency factor proposed at lowest frequency level. It can be predicted that natural frequency of the turbine blades occurs at lower range of frequency.

The bonded MFC type performs better which is corresponded by Tien and Goo (2010) reports. The MFC performance is depended on the distance of MFC layer from the neutral axis of the structure. Theoretically, the deformation is increased with the increment of the distance from the neutral axis. Comparison between embedded and bonded MFC is corresponded between experiment and statistical analysis. The bonded MFC is proposed as the best selection to harvest higher micro energy.

5. Structural health monitoring (SHM) identification for turbine blades via modal approach

This experiment was conducted to achieve the objective of a fundamental assessment of the SHM in kenaf wind turbines. The results are limited to pattern recognition between damaged and normal blades for bonded and embedded Macro Fiber Composite (MFC) as shown in Figure 17. The comparisons of frequency response function graphs between damaged and normal blades are exhibited in Figure 18 for bonded MFC and Figure 19 for embedded MFC.

Figure 17 shows the overall average magnitude (dBV) of bonded and embedded MFC in normal blades and damage blades. Generally, a normal turbine blade with bonded MFC illustrates the highest magnitude compared to damage turbine blades with bonded MFC. Similar to embedded MFC, normal blade exhibit a higher magnitude as compared to damage blades. This indicates that the SHM system can perform well and works in the kenaf wind turbine blade system. The findings in bonded MFC of normal blades support the previous experiment results which show the highest magnitude in induced electricity throughout the range of experiment. Embedded MFC exhibits a lower magnitude than bonded MFC. This report agreed with vibration tests on the turbine blade and verifies that bonded MFC performs better due to the significant length from the natural axis. Besides that, it shows that damage blade causes the magnitude response to reduce hence assisting in identifying the structural integrity of the blade and proves that SHM is feasible in the kenaf wind turbine blade system.

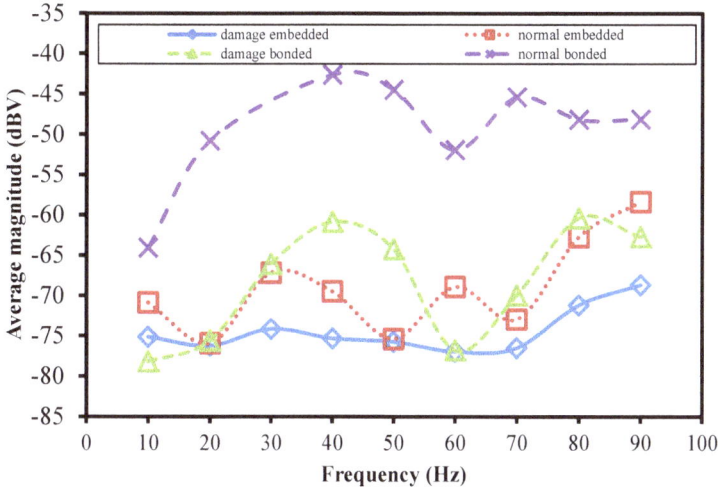

Figure 17 Comparison of Frequency Response Function graph between damage blade embedded MFC, normal blade embedded MFC, damage blade bonded MFC and normal blade bonded MFC

6. The comparison graph between damaged and normal blades for bonded MFC

Figure 17 presents a comparison of frequency response functions for bonded MFC between damaged and normal blade at a different exciter frequency. The normal blades result is presented in blue colour whilst the damage blade is illustrated in red colour. Generally, the normal blade shows a higher magnitude as compared to the damaged blade in all figures. This behaviour shows that damage and normal condition can be predicted and determined via bonded and embedded MFC especially in kenaf turbine blade application at exciter frequency range from 10Hz to 90Hz. Hence, it may assist the engineer to observe and monitor the structural integrity of the wind turbine system.

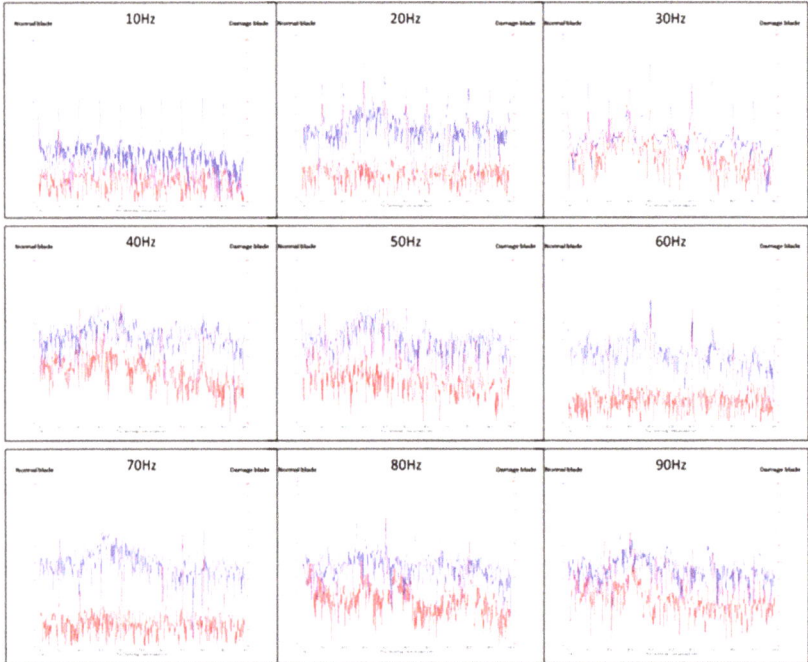

Figure 18 Comparison of Frequency Response Function graph between normal and damage blade from 10Hz to 90Hz excitation from shaker in bonded MFC

The graph peak in the frequency response function is clearly illustrated in the graph at the exciter frequency of 10, 20, and 30 Hz. Clear peak lines disappear in other exciter frequencies. The noise from the ambient response and higher exciter frequencies may influence the emerging of peak points in each graph line. Furthermore, the peak lines for each graph for damage blades is decreased compared to the normal graph. It may be due to changes in natural frequency of the turbine blade as the weight is reduced.

Fundamentally, the SHM system can be employed via bonded MFC in the wind turbine system and further analysis should be conducted to systematically understand the behaviour such as critical point, optimum layer thickness of blade and MFC location for better performance. Furthermore, the effect of embedded technique should be explored as well and will be presented in the next section.

7.   Comparison graph between damage and normal blade for embedded MFC

The comparison of frequency response function graph for embedded MFC between damaged and normal blades is reported in Figure 19. The normal blade result is exhibited in blue colour whilst the damage blade is illustrated in red colour. Both normal and damaged graph line show a similar trend and overlap for most exciter frequencies except for 30, 60, 80 and 90 Hz. The graph line in bonded MFC shows a clear difference between normal and damaged blades but the difference in embedded MFC is small. However, as in the bonded MFC analysis, the damaged blade frequencies magnitude is reduced compared to the normal blade. Several factors may influence the frequency magnitude such as its location from the neutral axis, embedded MFC and damage distance.

On the other hand, peak point analysis shows that all the graphs from each exciter frequency exhibits a clear peak line except for 80 and 90 Hz exciter frequency. The normal blade illustrates a clear peak as compared to the damage blades. From the observation and above discussion, embedded MFC demonstrates low sensitivity to the damage of the structure as the frequency magnitude for normal and damaged blade shows almost the same trend and at some points an overlap occurred. However, the peak trend in normal blade of embedded MFC compared to bonded MFC shows that it responds highly with the deformation due to external forces exerted on the turbine blade.

Structural Health Monitoring system in kenaf wind turbines is a feasible technique based on the experiments conducted. The signal basic behaviour for embedded and bonded MFC are recorded but need further investigation. Bonded MFC perform well in differentiate the damage blade and normal blade signal as compared to embedded MFC. Whilst, embedded MFC shows high sensitivity with the external forces apply to the turbine blade.

Further investigation needs to be conducted to understand several factors influencing the system performance for structural health monitoring application such as the natural frequency behaviour of the structure, ambient effect to the signal noise, critical point of the turbine structure, optimum layer thickness of blades, MFC locations for better performance and damage distance.

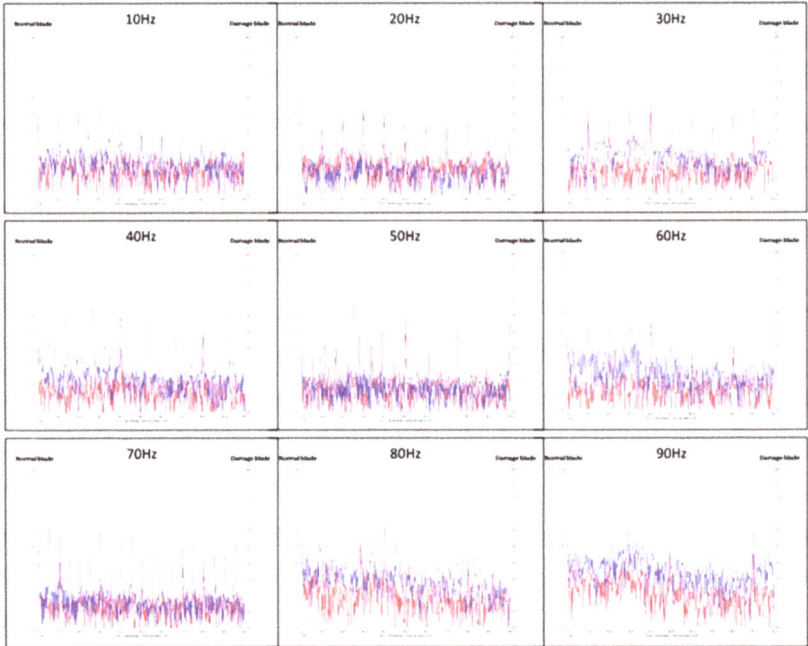

Figure 19 Comparison of Frequency Response Function graph between normal and damage blade from 10Hz to 90Hz excitation from shaker in embedded MFC

8. Summary

Application of woven kenaf in wind turbine blade opens a new investigation especially on micro energy harvester by MFC. The MFC bonded factor is deeply investigated to understand the behavior influenced the performance of micro energy harvester. Besides that, application of SHM in biocomposite wind turbine blade is preliminary explored for future investigation.

Wind turbine micro energy harvester testing and Taguchi analysis present that bonded technique suggested as the most influenced factor in micro energy harvesting at the vibration range of 10 to 90 Hz. Hence, bonded to the surface is proposed as the best technique in promoting higher micro energy harvesting for woven kenaf turbine blades at the vibration range of 10 to 90 Hz. Besides that, SHM system is practically can be applied simultaneously with micro energy harvesting system in turbine blades.

Acknowledgements:

The authors would like to thanks Ministry of Education, Malaysia for providing Mybrain 15 grant (MyPhD) and Universiti Putra Malaysia for providing the research grants FRGS 5524501 and RUGS 978100 for this research work.

References

[1]   Ali, W. G. and S. W. Ibrahim (2012). "Power Analysis for Piezoelectric Energy Harvester." Energy and Power Engineering 4: 496-505.

https://doi.org/10.4236/epe.2012.46063

[2]   Anton, S. R. and D. J. Inman (2008). Energy harvesting for unmanned aerial vehicles. Student Research Conference. Old Dominion University, Norfolk, Virginia, Virginia Space Grant Consortium.

https://doi.org/10.1117/12.774990

[3]   Anton, S. R. and H. A. Sodano (2007). "A review of power harvesting using piezoelectric materials (2003–2006)." Smart materials and structures 16: R1-R21.

https://doi.org/10.1088/0964-1726/16/3/R01

[4]   Anton, S. R., S. G. Taylor, et al. (2012). Multi-source energy harvesting for wind turbine structural healt monitoring node. Advances in Structural Health Management and Composite Structures, Jeonju, Republic of Korea.

[5]   Aslam Bhutta, M. M., N. Hayat, et al. (2012). "Vertical axis wind turbine – A review of various configurations and design techniques." Renewable and Sustainable Energy Reviews 16 1926– 1939.

https://doi.org/10.1016/j.rser.2011.12.004

[6]   Chong, W. T., A. Fazlizan, et al. (2012). "Early development of an innovative building integrated wind, solar and rain water harvester for urban high rise application." Energy and Buildings 47: 201–207.

https://doi.org/10.1016/j.enbuild.2011.11.041

[7]   Chong, W. T., M. S. Naghavi, et al. (2011). "Techno-economic analysis of a wind–solar hybrid renewable energy system with rainwater collection feature for urban high-rise application." Applied Energy 88: 4067-4077.

https://doi.org/10.1016/j.apenergy.2011.04.042

[8]   Chong, W. T., K. C. Pan, et al. (2013). "Performance investigation of a power augmented vertical axis wind turbine for urban high-rise application." Renewable Energy 51: 388-397.

https://doi.org/10.1016/j.renene.2012.09.033

[9]   Chong, W. T., S. C. Poh, et al. (2012). "Vertical axis wind turbine with omni-directional guide vane for urban high-rise buildings." J. Cent. South Univ. 19: 727-732.

https://doi.org/10.1007/s11771-012-1064-8

[10] Ciang, C. C., J.-R. Lee, et al. (2008). "Structural health monitoring for a wind turbine system: a review of damage detection methods." Measurement Science & Technology 19(12).

https://doi.org/10.1088/0957-0233/19/12/122001

[11] Dabiri, J. o. (2011). "Potential order-of-magnitude enhancement of wind farm power density via counter-rotating vertical-axis wind turbine arrays." Journal of Renewable and Sustainable Energy 3.

https://doi.org/10.1063/1.3608170

[12] Elfrink, R., T. M. Kamel, et al. (2009). "Vibration energy harvesting with aluminum nitride-based piezoelectric devices." Journal of Micromechanics and MicroEngineering 19(9): Paper No. 094005.

[13] Erturk, A., O. Bilgen, et al. (2008). "Power generation and shunt damping performance of a single crystal lead magnesium niobate-lead zirconate titanate unimorph: Analysis and Experiment." Applied Physics Letters 93(22).

https://doi.org/10.1063/1.3040011

[14] Erturk, A. and D. J. Inman (2009). "An experimentally validated bimorph cantilever model for piezoelectric energy harvesting from base excitations." Smart Materials and Structures 18(2).

https://doi.org/10.1088/0964-1726/18/2/025009

[15] Farinholt, K. M., N. A. Pedrazas, et al. (2009). "An Energy Harvesting Comparison of Piezoelectric and Ionically Conductive Polymers." Journal of Intelligent Material Systems and Structures 20: 633-642.

https://doi.org/10.1177/1045389X08099604

[16]   Flynn, A. M. and S. R. Sanders (2002). "Fundamental limits on energy transfer and circuit considerations for piezoelectric transformers." IEEE Transaction on Power Electronics 17(1): 8-14.

https://doi.org/10.1109/63.988662

[17]   Gavald, J., J. Massons, et al. (1990). "Experimental study on a self-adapting darrieus-savonius wind machine." Solar & Wind Tecnology 7(4): 457-461.

https://doi.org/10.1016/0741-983X(90)90030-6

[18]   Goldschmidtboeing, F. and P. Woias (2008). "Characterization of different beam shapes for piezoelectric energy harvesting." Journal of Micromachining and Microengineering 18: 104013.

https://doi.org/10.1088/0960-1317/18/10/104013

[19]   Greenblatt, D., M. Schulman, et al. (2012). "Vertical axis wind turbine performance enhancement using plasma actuators." Renewable Energy 37: 345-354.

https://doi.org/10.1016/j.renene.2011.06.040

[20]   Herbert, G. M. J., S. Iniyan, et al. (2007). "A review of wind energy technologies." Renewable and Sustainable Energy Reviews 11: 1117–1145.

https://doi.org/10.1016/j.rser.2005.08.004

[21]   Heung, S. K., K. Joo-Hyong, et al. (2011). "A Review of Piezoelectric Energy Harvesting Based on Vibration." International Journal of Precision Engineering and Manufacturing 12(6): 1129-1141.

https://doi.org/10.1007/s12541-011-0151-3

[22]   Hossain, A., A. K. M. P. Iqbal, et al. (2007). "Design and development of a 1/3 scale vertical axis ind turbine for electrical power generation." Journal of Urban and Environmental Engineering 1(2): 53–60.

https://doi.org/10.4090/juee.2007.v1n2.053060

[23]   Howell, R., N. Qin, et al. (2010). "Wind tunnel and numerical study of a small vertical axis wind turbine." Renewable Energy 35: 412–22.

https://doi.org/10.1016/j.renene.2009.07.025

[24]   Hyun Jeong, S., Y.-T. Choi, et al. (2010). "Energy Harvesting Devices Using Macro-fiber Composite Materials." Journal of Intelligent Material Systems and Structures 21(6): 647-658.

https://doi.org/10.1177/1045389X10361633

[25]   Ibrahim, A.-B. (2009). "Building a wind turbine for rural home." Energy for Sustainable Development 13 159–165.

https://doi.org/10.1016/j.esd.2009.06.005

[26]   Islam, M., M. R. Amin, et al. (2007). Aerodynamic factors affecting performance of straight-bladed vertical axis wind turbines. ASME international mechanical engineering congress and exposition.

https://doi.org/10.1115/imece2007-41346

[27]   Islam, M., D. S. K. Ting, et al. (2008). "Aerodynamic models for Darrieus-type straight-bladed vertical axis wind turbines." Renewable and Sustainable Energy Reviews 12: 1087-1109.

https://doi.org/10.1016/j.rser.2006.10.023

[28]   Jeong, S. J., D. S. Lee, et al. (2011). "Properties of piezoelectric ceramic with textured structure for energy harvesting." Ceramic International: 5-14.

[29]   Kanellos, F. D. and N. D. Hatziargyriou (2008). "Control of variable speed wind turbines in islanded mode of operation." Ieee Transactions on Energy Conversion 23(2): 535-543.

https://doi.org/10.1109/TEC.2008.921553

[30]   Li, L. (2012). Vibrations Analysis of Vertical Axis Wind Turbine. School of Engineering and Advanced Technology. New Zealand, Massey University. Master of Engineering.

[31]   Liu, H., S. Zhang, et al. (2012). "Development of piezoelectric microcantilever flow sensor with wind-driven energy harvesting capability." Applied physics letters 100: 223905.

https://doi.org/10.1063/1.4723846

[32]   Mertens, S., G. van-Kuik, et al. (2003). "Performance of an H-Darrieus in the skewed flow on a roof." Journal of Solar Energy Engineering 125: 433–41.

https://doi.org/10.1115/1.1629309

[33]   Mohamed, M. H. (2012). "Performance investigation of H-rotor Darrieus turbine with new airfoil shapes." Energy 47: 522-530.

https://doi.org/10.1016/j.energy.2012.08.044

[34] Park, K.-s., T. Asim, et al. (2012). "Computational Fluid Dynamics based Fault Simulations of a Vertical Axis Wind Turbines." Journal of Physics: Conference Series 364: 012138.

https://doi.org/10.1088/1742-6596/364/1/012138

[35] Saeidi, D., A. Sedaghat, et al. (2013). "Aerodynamic design and economical evaluation of site specific small vertical axis wind turbines." Applied Energy 101: 765–775.

https://doi.org/10.1016/j.apenergy.2012.07.047

[36] Sandra Eriksson, H. Bernhoff, et al. (2008). "Evaluation of different turbine concepts for wind power." Renewable and Sustainable Energy Reviews 12: 1419–1434.

https://doi.org/10.1016/j.rser.2006.05.017

[37] Shen, D., J. H. Park, et al. (2009). "Micromachined PZT cantilever based on SOI structure for low frequency vibration energy harvesting." Sensors and Actuators A: Physical 154(1): 103-108.

https://doi.org/10.1016/j.sna.2009.06.007

[38] Sodano, H. A. (2003). Macro-Fiber Composites for Sensing, Actuation and Power Generation. Faculty of the Virginia Polytechnic Institute and State University.

[39] Sodano, H. A., D. J. Inman, et al. (2004). "A review of power harvesting from vibration using piezoelectricmaterials." The Shock and Vibration Digest 36(3): 197-205.

https://doi.org/10.1177/0583102404043275

[40] Sodano, H. A., D. J. Inman, et al. (2005). "Comparison of piezoelectric energy harvesting devices for recharging batteries." Journal of Intelligent Material Systems and Structures 16(10): 799-807.

https://doi.org/10.1177/1045389X05056681

[41] Sodano, H. A., G. H. Park, et al. (2003). Electric power harvesting using piezoelectric materials. Center for Intelligent Material Systems and Structures, Virginia Polytechnic Institute and State University.

[42] Stein, P., M.-C. Hsu, et al. (2012). "Operator- and template-based modeling of solid geometry for Isogeometric Analysis with application to Vertical Axis Wind Turbine simulation." Comput. Methods Appl. Mech. Engrg. 213-216: 71–83.

https://doi.org/10.1016/j.cma.2011.11.008

[43]   Sukanta, R., B. Agnimitra, et al. (2010). CFD analysis of an airfoil shaped three bladed H-Darrious rotor made from fibreglass reinforced plastic (FRP). Proceedings of the 37th National & 4th International Conference on Fluid Mechanics and Fluid Power, IIT Madras, Chennai, India.

[44]   Sundaresan, M. J. and M. J. Schulz (August 2006). Smart Sensor System for Structural Condition Monitoring of Wind Turbines. Midwest Research Institute., National Renewable Energy Laboratory.

[45]   Sundaresan, M. J., M. J. Schulz, et al. (August 1999). Structural Health Monitoring Static Test of a Wind Turbine Blade. Midwest Research Institute, National Renewable Energy Laboratory.

[46]   Taguchi, G. (1990). Introduction to Quality Engineering. Tokyo, Asian Productivity Organization.

[47]   Tai, F.-Z., K.-W. Kang, et al. (2012). "Study on the analysis method for the vertical-axis wind turbines having Darrieus blades." Renewable Energy: 1-6.

[48]   Takao, M., H. Kuma, et al. (2009). "A straight-bladed vertical axis wind turbine with a directed guide vane row effect of guide vane geometry on the performance." Journal of Thermal Science 18: 54-7.

https://doi.org/10.1007/s11630-009-0054-0

[49]   Tien, C. M. T. and N. S. Goo (2010). "Use of a piezocomposite generating element in energy harvesting." Journal of Intelligent Material Systems and Structures 21(14): 1427-1436.

https://doi.org/10.1177/1045389X10381658

[50]   Vandenberghe, D. and E. Dick (1987). "A free vortex simulation method for the straight bladed vertical axis wind turbine." Journal of Wind Engineering and Industrial Aerodynamics 26: 307-324.

https://doi.org/10.1016/0167-6105(87)90002-X

[51]   Wang, Y. (2012). Simultaneous Energy Harvesting and Vibration Control via Piezoelectric Materials. Virginia Polytechnic Institute and State University. Blacksburg. Doctor of Philosophy: 138.

[52]   Yeh, T.-H. and L. Wang (2008). "A study on generator capacity for wind turbines under various tower heights and rated wind speeds using Weibull distribution." Ieee Transactions on Energy Conversion 23(2): 592-602.

https://doi.org/10.1109/TEC.2008.918626

[53]   Zhu, D., A. Almusallam, et al. (2010). A Bimorph Multi-layer Piezoelectric Vibration Energy Harvester. Proceedings of Power MEMS Belgium.

Keyword Index

About the Authors

F. Mustapha
Faizal Mustapha received Ph.D. degree in Structural Health Monitoring System from University of Sheffield, U.K, in 2006. Now he works as Associate Professor at Universiti Putra Malaysia, Malaysia. His current research interests include damage identification, sensor technology, advanced material, advanced signal processing.

A. Hamdan
Ahmad Hamdan Ariffin graduated Ph.D in Structural Health Monitoring System and Biocomposite wind turbine from Universiti Putra Malaysia, Malaysia in 2015. He received Master of Engineering Science degree in Advanced Manufacturing from University of Malaya, Malaysia in 2011. His current research interests include biocomposite, material characterization and manufacturing.

Nisreen N. Ali Al-Adnani
Nisreen N. Ali Al-Adnani graduated PhD in Civil Engineering from Universiti Putra Malaysia, Malaysia in 2014. Her research interests are in composite, Structural Health Monitoring and civil engineering

K.D. Mohd Aris
Khairul Dahri Mohd Aris graduated PhD in Composite Structure from Universiti Putra Malaysia, Malaysia in 2014. He works as a senior lecturer as well as a researcher in Malaysian Institute of Aviation Technology, University of Kuala Lumpur. Now, he is focusing in composite of aircraft structure, damage identification and aircraft repair technology.

www.ingramcontent.com/pod-product-compliance
Lightning Source LLC
Chambersburg PA
CBHW071229210326
41597CB00016B/2000